The Physics of the Interstellar Medium

The Physics of the Interstellar Medium

Third Edition

J E Dyson
D A Williams

CRC Press is an imprint of the
Taylor & Francis Group, an informa business

Front cover image: *The image was made by NASA's Spitzer Space Telescope operating in four wavebands in the infrared, between 3.6 and 8.0 micrometres, and shows star formation occurring in a dense globule of gas that is embedded in high pressure ionized gas surrounding a massive star to the left of the globule (not shown in this image). The star also has a powerful wind that together with the ionized gas compresses the globule, making a dense rim. The wind creates a long tail. Two young stars have been created within the dense gas in the head of the globule. Their winds have swept a spherical cavity in the gas of the globule.* Credit: NASA/JPL-Caltech/W. Reach (SSC/Caltech)

Third edition published 2021
by CRC Press
6000 Broken Sound Parkway NW, Suite 300, Boca Raton, FL 33487-2742

and by CRC Press
2 Park Square, Milton Park, Abingdon, Oxon, OX14 4RN

Second edition published by Taylor & Francis Group 1997

First edition published by Manchester University Press 1980

CRC Press is an imprint of Taylor & Francis Group, LLC

International Standard Book Number-13: 978-0-367-45732-7 (Hardback)
International Standard Book Number-13: 978-0-367-90423-4 (Paperback)
International Standard Book Number-13: 978-1-003-02503-0 (eBook)

Library of Congress Cataloging-in-Publication Data

Names: Dyson, J. E. (John Edward), 1941- author. | Williams, David A., 1937- author.
Title: The physics of the interstellar medium / J.E. Dyson & D.A. Williams.
Description: Third edition. | Boca Raton : CRC Press, 2020. | Includes bibliographical references and index.
Identifiers: LCCN 2020017203 | ISBN 9780367904234 (paperback) | ISBN 9780367457327 (hardback) | ISBN 9781003025030 (ebook)
Subjects: LCSH: Interstellar matter.
Classification: LCC QB790 .D97 2020 | DDC 523.1/125–dc23
LC record available at https://lccn.loc.gov/2020017203

Contents

Preface to the Third Edition

Our knowledge of the interstellar medium in the Milky Way galaxy – and in many external galaxies – has increased enormously since John Dyson and I wrote the first edition of this book, published in 1980. Our understanding of the importance of this tenuous component in driving the evolution of all galaxies has developed very substantially since the second edition was published in 1997. However, the main purpose of this book remains unchanged: we seek to illustrate many interesting aspects of physics by describing some unfamiliar situations that arise in interstellar space. This third edition of the book is aimed primarily at undergraduates on physics courses, as were the earlier editions. Although the book does not give a comprehensive coverage of the interstellar medium, it has nevertheless also been used by other readers as a concise introduction to the study of the interstellar medium, and I hope that the new edition will also be useful for this purpose.

Perhaps the most important area in which studies of the interstellar medium have developed is in our understanding of the formation of stars and planets occurring within very dense and dusty regions of the interstellar medium. This work depends to a large extent on the detection of emissions from molecules and dust within the dense interstellar gas. Infrared emissions from dust and molecules are not only important tracers of the gas in regions that are opaque to optical radiation, but they also are significant players in controlling the physical conditions and evolution of gas in star-forming regions. In this third edition, more prominence is given to the interstellar chemistry that generates interstellar molecules, to the properties of interstellar dust, and to the grand narrative that describes current ideas on how the formation of stars and planets occurs in the interstellar medium.

The original structure of the book is retained in this third edition, and the intellectual level required by a reader and the aim to present brief, simple explanations in a short book are maintained as in earlier editions. The opportunity is taken to include some recent astronomical images within the chapter texts, where appropriate. I am grateful for advice on some aspects of the revision from friends and colleagues, especially Cesare Cecchi-Pestellini and Serena Viti, but all errors and omissions are mine.

David A Williams

Some Relevant Physical and Astronomical Information

ENERGY CONVERSION TABLE

	Joule	Electron volt	Kelvin
Joule	–	6.2414×10^{18}	7.2430×10^{22}
Electron volt	1.6022×10^{-19}	–	1.1605×10^{4}
Kelvin	1.3806×10^{-23}	8.6173×10^{-5}	–

SOME USEFUL PHYSICAL CONSTANTS

Velocity of light, c	2.9979×10^{8} m s^{-1}
Gravitational constant, G	6.6743×10^{-11} m^{3} kg^{-1} s^{-2}
Planck constant, h	6.6261×10^{-34} J s
Dirac constant, $\hbar = h/2\pi$	1.05457×10^{-34} J s
Electron charge, e	1.60218×10^{-19} C
Electron rest mass, m_e	9.10938×10^{-31} kg
Proton rest mass, m_p	1.67262×10^{-27} kg
Boltzmann constant, k	1.38065×10^{-23} J K^{-1}
Avogadro number, N_A	6.02214×10^{23} mol^{-1}
Bohr radius, a_0	0.529177×10^{-10} m
Permittivity of free space, ε_0	8.85419×10^{-12} C^{2} N^{-1} m^{-2}

ASTRONOMICAL DATA

Time and distance measurements

I year	3.1557×10^{7} s
1 light year (ly)	9.461×10^{15} m
1 parsec (pc)	3.086×10^{16} m = 3.26 ly

SOLAR DATA

Mean distance from Earth, 1 astronomical unit (1 au) = 1.496×10^8 km = 8.3 light minutes.

Solar mass	1.989×10^{30} kg
Solar radius	6.957×10^8 m
Solar mean density	1408 kg m^{-3}
Solar age	4.6×10^9 yr
Solar energy emission rate	3.828×10^{26} J s^{-1}

MILKY WAY GALAXY DATA

Diameter	~30 kpc
Mass	~10^{12} solar masses
Radius at the Sun	~8 kpc
Rotation period at the Sun	2.4×10^8 yr
Age of the Milky Way galaxy	1.2×10^{10} yr

STELLAR CLASSIFICATIONS

Very occasionally in this book we refer to a type of star using a conventional classification system. The system is based on lines that are observed in the stellar spectrum which are used to determine the effective temperature and other properties of the star. Stars are arranged in classes O, B, A, F, G, K, and M, where O stars are the hottest (effective temperatures greater than about 30 000 K) and M stars are coolest (around 2400–3700 K). The Sun is a G star (temperature 5200–6000 K). These broad classes are subdivided by a number running from 0 (hottest) to 9 (coolest), so for example, the Sun is a G2 star. There are several other descriptors that refer to luminosity, evolutionary state, and spectral peculiarities.

Authors

John Dyson made outstanding research contributions over many years to our understanding of the responses of interstellar media to winds from stars and from active galaxies. He had a huge influence on these subjects and his work gained an international reputation. Much of his career was at the University of Manchester where he became Professor of Astronomy and Head of Astrophysics. He moved in 1996 to the University of Leeds, becoming Dean of Research, and was appointed Emeritus Research Professor in 2006.

He died in 2010 and is much missed by friends and colleagues world-wide who valued his scientific insight, quick wit, kindness and generosity.

David Williams is currently Emeritus Perren Professor of Astronomy at University College London. While at NASA's Goddard Space Flight Center in the 1960s he became interested in interstellar molecules and interstellar dust as potential probes of the interstellar medium. When John Dyson and David were both working in Manchester, John emphasised the importance of cosmic gas dynamics in understanding interstellar chemistry and dust, and David built a research group at UMIST to investigate these and other topics. He left Manchester in 1994 for UCL and has continued to study problems in interstellar physics and chemistry.

1 Introduction

1.1 GALAXIES AND THE GALAXY

If we are fortunate enough to see the sky on a moonless and cloudless night, then, when our eyes have become accustomed to the dark, we can discern the prominent feature called the Milky Way. This is a band across the sky containing a large number of bright stars and a background of many fainter stars – so many, in fact, that our eyes cannot resolve them and we see their light as a diffuse (i.e. 'milky') luminosity. This band of stars appears across the sky because the Sun is one of a vast collection of stars, most of which are arranged in the shape of a disc. When we look in the plane of the disc, we see many stars; when we look out of the plane, we see fewer stars. This collection of stars – most of them in the plane of the disc and some out of the plane in a spherical region – is called the Milky Way galaxy. The Milky Way is only one galaxy among an enormous number of galaxies.

The shape and extent of a galaxy, seen with the eye or a telescope, are defined by the stars it contains. This book, however, is concerned with material in the Milky Way galaxy, material that our eyes do not easily see but which, nevertheless, is present and which plays an essential role in the evolution of the galaxy. This material is the *interstellar medium.* In this book, we'll describe some of the interesting applications of physics to the study of the interstellar medium of the Milky Way galaxy, and at the same time, we'll try to explore the fundamental role it plays. However, this is a book about the *physics* of the interstellar medium in the Milky Way and should be read as a simple account of some applications of physics. It isn't a comprehensive description of all aspects of the interstellar medium of the Milky Way galaxy.

Before we come to describe evidence for the existence of material in the space between the stars, we should first have some idea of the dimensions of the galaxy. Our Sun appears to be a fairly typical star of the Milky Way galaxy, which is estimated to contain about 10^{11} stars. The Sun is not located in any special position in the galaxy, such as the galactic centre, but is in the plane of the disc about two-thirds of the galactic radius from the galactic centre. We cannot see, either by eye or by optical telescope, more than a mere fraction of the galaxy, because the intensity of visible light is diminished by a general extinction as the light travels in the plane of the galaxy. Out-of-plane directions are less extinguished. We shall have more to say about that interstellar extinction, but it was the first sign that the interstellar medium wasn't empty: evidently, it contains something that absorbs and scatters starlight. However, as we shall see, the galaxy can be investigated by other means, and we know that its many stars are distributed over a large volume. The diameter of the disc is of the order of 30 kpc (or about 100 thousand light years), and its thickness is about 2 kpc (or about 6 thousand light years). Stars are therefore about 10 light years from their nearest neighbours, on average, though the distribution of stars in space is far from uniform. Obviously, stars are very small compared with these dimensions, so that essentially all the vast space inside the galactic volume is occupied by the interstellar medium, which – as we'll describe – is a non-uniformly distributed dusty gas. The whole galaxy is gravitationally bound; it has spiral structure within it, and it rotates. The shape and structure of the Milky Way galaxy are thought to be similar to those of some external galaxies, and an image of a galaxy believed to be similar to the Milky Way is shown in Figure 1.1.

FIGURE 1.1 Image of spiral galaxy M74. This galaxy is seen by face-on by observers on Earth. It is about 10 Mpc distant, contains about 10^{11} stars, is slightly smaller than the Milky Way, and shares similar spiral structure. The arms contain many bright blue stars and regions of red emission from ionized hydrogen. (Image credit: NASA, ESA, and the Hubble Heritage (STSC/AURA)-ESA/Hubble Collaboration; Acknowledgment: R Chandar (University of Toledo) and J Miller (University of Michigan).)

The Milky Way galaxy is merely one of an enormous number, perhaps ~10^{11}, of galaxies in the Universe, and the dimensions of the visible Universe are about a factor of about 10^5 larger than those of the Galaxy. We can sometimes investigate the properties of our own Galaxy by studying other galaxies that we can see more readily. The distant galaxies appear to be receding from our own Galaxy, with velocities increasing (and apparently accelerating) with distance. Distances between neighbouring galaxies are very much greater than the dimensions of galaxies themselves, and are typically measured in Mpc (or millions of light years). Therefore, the light gathered by our telescopes and focused to form an image of distant galaxies must have been travelling on its journey long before human beings evolved on Earth. The space between the galaxies – intergalactic space – is not our concern in this book, but, in passing, we note that any gas in intergalactic space must be much less dense than the gas in interstellar space. Studies show that galaxies exist in various forms. Some are irregular, some show a more well-defined disc shape, while some of these have spiral structure (spiral arms) within them. Some are classified as elliptical. Our own Galaxy is known to have spiral structure similar to M74. Figure 1.2 shows photographs of several types of galaxy. All galaxies have interstellar matter; they may be poorer or richer in interstellar matter than the Milky Way. Our discussion in this book is directed towards interstellar matter in the Milky Way galaxy, but the ideas expressed have general application to interstellar matter in all galaxies.

When we look at the sky, it is obvious that some stars appear much brighter than others. This is often because such stars are relatively nearby, but some stars appear bright because they are intrinsically powerful sources of radiation. Astronomers can, by a variety of means, deduce the masses and intrinsic luminosities of stars, and these results agree well with theories of stellar evolution. These theories tell us that stars have masses within a range of about 0.1 to about 100 times the mass of the Sun ($M_{\odot}$), and that luminosities corresponding to these masses may range from about 10^{-3} to 10^6 times the luminosity of the Sun. The more massive the star, the greater is its luminosity. Therefore, the brightest stars, containing nearly 100 times the amount of fuel as the Sun, are squandering it at a rate 10^6 times faster. We therefore expect that they can exist only for 10^{-4} times the life of the Sun, that is, for about a few million years. This seems a long time for us on Earth, but for the galaxy, these bright stars are transient objects, like candles which soon burn down. We are forced to conclude that such stars have formed in the recent past, and – by implication – that they are certainly forming now. The galaxy *evolves*; it was not formed in the state in which we see it now.

At the end of its life, a sufficiently massive star explodes violently in a dramatic event called a supernova. These explosions are so powerful that we can observe their effects even when they occur in distant galaxies; the exploding star may become (very briefly) as bright as its host galaxy. The last one known to occur in the Milky Way occurred more than 300 years ago, but on average the Milky Way may be host to two supernovae per century. Perhaps the most famous supernova that occurred in our own galaxy is the one that was observed in 1054 AD, and which caused the Crab nebula, an extended supernova remnant

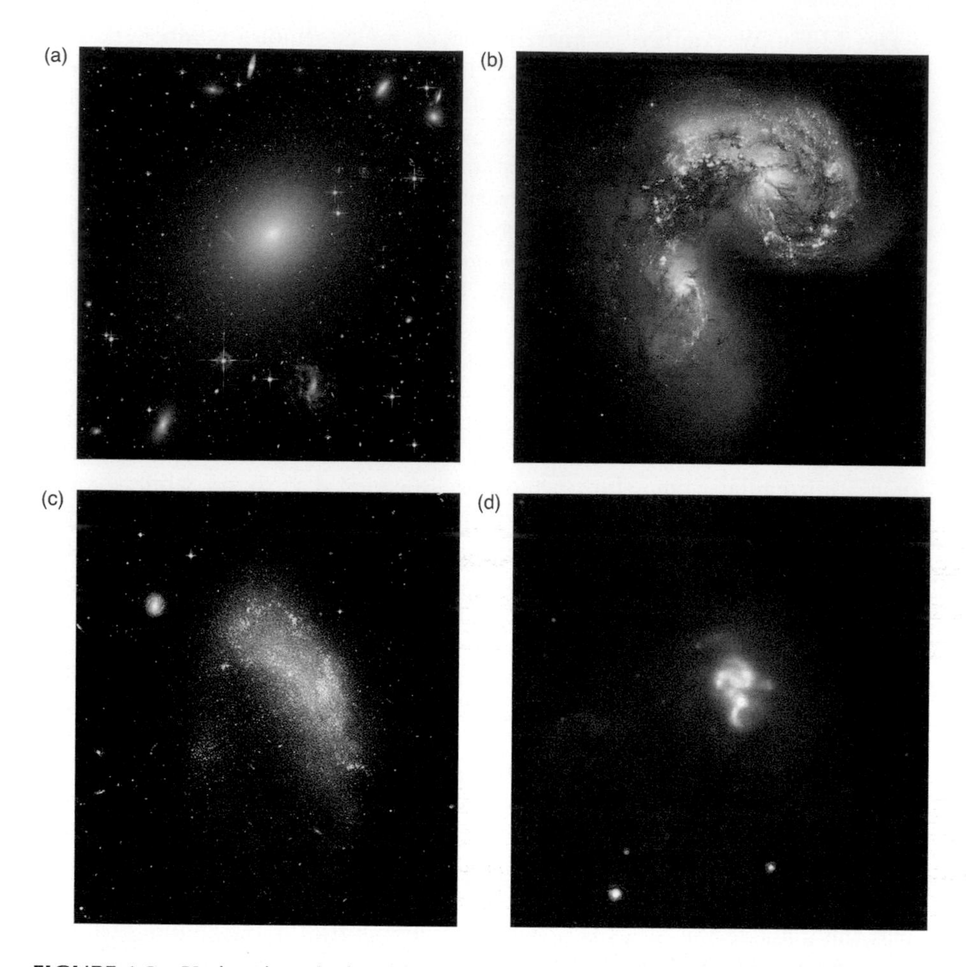

FIGURE 1.2 Variety in galaxies. (a) Elliptical galaxies, of which ESO 325-G004 is one example, have a smooth profile and an ellipsoidal shape. They have relatively few high-mass bright stars and consist mainly of low mass stars. (Credit: NASA, ESA, and The Hubble Heritage team and J Blakeslee.) (b) Starburst galaxies have a very high rate of star formation which can be stimulated by a close collision between galaxies. This is the case for the Antenna Galaxies, a merger between two galaxies – NGC 4038 and 4039. (Credit: ESA/Hubble and NASA.) (c) Irregular galaxies are often small and their shapes are a result of near collisions with other massive galaxies. The example shown is NGC 1427A. (Credit: NASA, ESA, and the Hubble Heritage Team.) (d) Some galaxies are very bright, especially in the infrared, and are known as luminous infrared galaxies, or ultraluminous infrared galaxies (ULIRG). IRAS 1927–0406 shown here is an example of a ULIRG. (Credit: NASA).

that we can still observe (Figure 1.3(a)). A more evolved supernova remnant is shown in Figure 1.3(b). It is in the Large Magellanic Cloud, a neighbouring galaxy to the Milky Way. In 1987, another supernova occurred in the Large Magellanic Cloud. It (SN1987A) has become the best studied of all supernovae.

Supernova explosions eject material in large amounts, comparable to the mass of our Sun, from the interiors of stars into interstellar space. This ejected material is rich in the 'ashes' of the thermonuclear processes which power the stars, and these 'ashes' are the elements heavier than hydrogen. Less-massive stars also contribute to the enrichment of the interstellar medium with the 'ashes' of thermonuclear burning, albeit in a less dramatic way. The fact that we and our Earth are made predominantly from these elements, and that the Sun also contains them, indicates that the Solar System is made from 'recycled' material, that is, it was formed from interstellar gases containing the 'ashes' of a previous generation of stars. The picture that emerges from these general considerations is that as the galaxy evolves, stars are formed, and they 'age' and 'die', sometimes explosively. The ejected material enriches the interstellar medium with atoms of elements other than hydrogen, and the gases in interstellar space somehow condense to form a new generation of stars, and the cycle is repeated. In this scenario, we see that the interstellar medium plays a crucial role: it is the reservoir of mass for forming new stars and planets in any galaxy, even though it may at first seem a minor component of a galaxy, and it is continually enriched in the heavy elements that are the ashes of thermonuclear processes in stars.

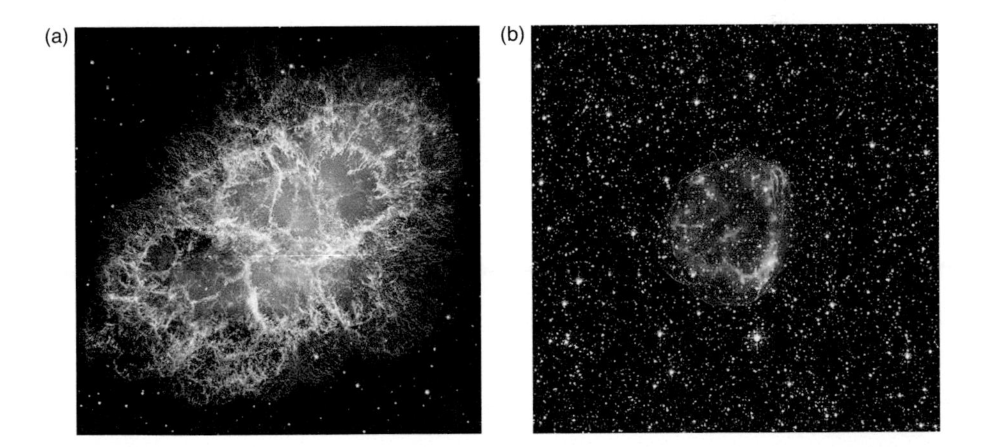

FIGURE 1.3 (a) The supernova remnant known as the Crab nebula. (Credit: NASA, ESA, J Hester, and A Loll (Arizona State University).) (b) An image of the supernova remnant SNR 0519690 which represents X-ray emission from very hot gas in the nebula as blue (Chandra Observatory) and optical emission in red (Hubble Space Telescope) from the outer boundaries of the nebula. (Credit: X-ray; NASA/CXC/Rutgers/J Hughes: Optical; NASA/STScI.)

1.2 EVIDENCE FOR MATTER BETWEEN THE STARS

Naked-eye observation of the Galaxy tells us nothing about interstellar space – except that it appears to be empty. However, this is easily seen to be an incorrect conclusion. Photographs from even low-powered telescopes provide the most striking and visually beautiful evidence of matter in interstellar space. Where interstellar gas is so near to a star that it is hot, it will radiate, and we can detect this radiation and identify it. We show later in this book images of some of these radiating regions.

The main evidence for the existence of interstellar matter rests, as we shall see in the succeeding chapters, on spectroscopy. The atoms, ions, and molecules in the interstellar gas may emit or absorb radiation corresponding to transitions between their various energy levels. A well-known transition is that of hydrogen atoms giving rise to the 21 cm radio line. Hydrogen atoms may also give rise to many other lines; the one giving the red colour prominent in many photographs of nebulae corresponds to the transition H(3p) $\rightarrow$ H(2s) at 656.3 nm. There are also other ways in which emission of radiation can occur, in particular over a continuum of wavelengths. All such radiation carries with it information concerning the interstellar medium, as we shall discuss in the following chapter.

Gravitational effects within the Galaxy also give indirect evidence for the existence of interstellar matter, and allow us to deduce an upper limit to the interstellar density. The argument goes as follows: a star situated at a distance z above or below the galactic plane experiences an acceleration, g_z, towards the plane due to the gravitational interaction of the star with all the mass in the Galaxy, stars plus gas. We cannot measure this acceleration for a single star. However, it can be deduced from the behaviour of a population of stars at various distances, z, from the plane, and so we may find g_z as a function of z. The total mass in the plane has a mean density, $\bar{\rho}$, which is related in a simple way to the rate of change of g_z with z, dg_z/dz, so $\bar{\rho}$ may be calculated. This mean density is found to be about 10^{-20} kg m^{-3} at the mid-plane. The known stars contribute to this mean density about 4×10^{-21} kg m^{-3}, so that the upper limit to the mass density of unobserved faint stars, other condensed objects, and the interstellar matter is 6×10^{-21} kg m^{-3}, or equivalent to about 2.7×10^{6} hydrogen atoms m^{-3}, allowing for 10% (by number) of the atoms being helium.

The centre of the Milky Way galaxy cannot be seen at optical wavelengths but is revealed at infrared wavelengths, see Figure 1.4.

1.3 PREVIEW

In this book, we are concerned with the physics of the interstellar medium in the Milky Way galaxy, but we shall particularly bear in mind the intimate relation between interstellar matter and galactic evolution through star and planet formation. The picture outlined in Section 1.1 – in which gas clouds collapse to form stars which themselves age and die, sometimes explosively – gives a direction to the chapters.

FIGURE 1.4 The centre of the Milky Way, about 8 kpc distant from the Sun, cannot be observed at optical wavelengths because of interstellar extinction along that path. This false-colour image in the infrared (data from NASA's Spitzer Space Telescope) gives a clear view of the centre, showing emission from warm dust (in red), the presence of enormous numbers of cool stars (in blue), and an intense radio source containing a supermassive black hole (in white). (Credit: NASA/JPL-Caltech/S Stolovy (Spitzer Science Center/Caltech).)

We begin by asking how astronomers obtain information about the interstellar medium (Chapter 2). What are the physical principles behind the formation of a spectral line? What other processes cause the emission and absorption of interstellar radiation? We end the chapter with a description of the broad types of interstellar regions that have been identified. Chapter 3 describes processes occurring between atoms, ions, and molecules and the effects these have on the interstellar gas. A rich chemistry arises from these gas phase processes, and the molecules formed provide probes of a wide variety of physical conditions. Chapter 4 is concerned with interstellar dust grains, and we shall discuss some of the interesting physics involved in describing these solid particles and the effects they have in the interstellar medium.

Chapter 5 gives special attention to those regions which are visually impressive in astronomical photographs, that is, those regions of space near hot stars where the gas is ionized and is emitting copious amounts of radiation. Chapter 6 gives an introduction to gas dynamics, which enables us in Chapter 7 to discuss some of the dramatic dynamical events occurring in interstellar space: for example, the expansion of ionized nebulae into cool gas, the effects of high-speed stellar winds on interstellar gas, and – most dramatically of all – supernova explosions. The final chapter looks at some of the physics that is fundamental to the problem of star formation. In this process, part of a gas cloud of density, say, 10^9 H_2 molecules m^{-3} must eventually become a star with a mean density of, say, 10^{30} H nuclei m^{-3}. This is an enormous transformation.

1.4 UNITS

Astronomy requires a large number of special units to describe in a convenient manner the quantities it measures or calculates. Thus, the solar mass, $M_\odot$, and solar luminosity, $L_\odot$, are convenient measures of the mass and luminosity of other stars, rather than the kilogram and candela, respectively. When measuring distance, it is more appropriate to use a unit such as a light year or a parsec. In general, however, the book is written in SI units, and any special units needed are defined in the tables on pp ix and x. An exception to standard SI usage is the adoption of the electron volt (eV) in describing energy levels within atoms and molecules, and multiples of eV (e.g. MeV) in describing the energies of cosmic rays. Temperature may conveniently be regarded as an energy measure. The table on p ix gives conversion factors for the energy units used here.

The SI units used here are for the convenience of physics undergraduates. If, as we hope, they refer to some of the astrophysical literature, they will find that an older and more traditional system is generally in use by astrophysicists. We sympathize with the more experienced readers, who will find familiar quantities appearing here with unfamiliar values.

2 How We Obtain Information about the Interstellar Medium

2.1 INTRODUCTION

Astronomy advances by observation. We have to take what we are given: this is mainly (though not entirely) information encoded in the electromagnetic radiation we receive from the Universe, over the whole spectrum from highly energetic γ-rays to long-wavelength radio waves. In addition to electromagnetic spectra, there are some other carriers of astronomical information. We can detect energetic cosmic ray particles; these are highly energetic nuclei of atoms (mostly hydrogen nuclei) and highly energetic electrons. Cosmic ray particles may be detected either directly or indirectly by their effects on the Earth's atmosphere. Neutrinos from the Sun and other stars also flood the Earth and carry information about nuclear processes in stellar interiors; they can be detected on Earth in deeply shielded experiments. Gravitational waves have recently been detected in experiments of exquisite sensitivity; these waves arise from the merger of a pair of black holes. Meteorites and samples of moon rocks and of material from asteroids and comets in the solar system are the only macroscopic material from the extraterrestrial Universe which we are able to collect and then analyse in the laboratory. This chapter, however, concentrates on how we find information about the interstellar medium of our Milky Way galaxy and of other galaxies, from electromagnetic spectra.

2.2 SPECTRAL LINES

2.2.1 INTRODUCTION

The information carried by the electromagnetic spectrum may arise either in the formation of spectral lines, that is, in absorption or emission over a relatively narrow band of frequency, or in continuous emission or absorption over a wide band of frequencies. For example, radiation at wavelengths shorter than λ = 91.2 nm will suffer continuous absorption in a hydrogen atom gas. But in the vicinity of λ = 121.6 nm, there is absorption in a line. The continuous absorption arises because of the ionization of the H atom by the photon of energy $h\nu$ (or $\hbar\omega$; $\nu = c/\lambda$ is the frequency, and $\omega = 2\pi\nu$ is the angular frequency) $h\nu + \mathrm{H(1s)} \rightarrow \mathrm{H}^+ + e \quad (h\nu \geq 13.6\mathrm{eV}, \mathrm{or}\ \lambda \leq 91.2\mathrm{nm})$

and the line absorption arises because of the excitation

$$h\nu + \mathrm{H(1s)} \rightarrow \mathrm{H(2p)} \qquad (h\nu = 10.2\ \mathrm{eV}, \text{or } \lambda = 121.6\ \mathrm{nm}).$$

Lines may be observed either in emission or absorption. A line appears in emission if the sources of the line – atoms, molecules, or ions – have been excited by some means, possibly in collisions or by absorption of radiation. A line is seen in absorption if the source is in its lower state and absorbs from a background continuum, which may be stellar radiation or some other source. Lines are formed by many different types of transitions, some of which we discuss in Section 2.2.4. We shall refer in what follows to atoms, but what we say also applies to molecules and ions. The first problem to understand is why a line may sometimes be in absorption and sometimes in emission, and also why a line has a particular profile, or shape, that is, a particular plot of intensity versus frequency across the spectral line.

2.2.2 Line Shapes

Natural line shape. Emitting atoms produce lines of finite width, as we can see from the following classical argument. We regard the emitting atom as an oscillator, which is almost harmonic but which is lightly damped by its interaction with the emitted radiation. Then the equation of motion for this system is

$$m\frac{\mathrm{d}^2\boldsymbol{r}}{\mathrm{d}t^2} = m\ddot{\boldsymbol{r}} = -m\omega_0^2\boldsymbol{r} + \boldsymbol{F} \tag{2.1}$$

in which $\boldsymbol{r}$ is the position vector and $\boldsymbol{F}$ is the damping force. The oscillator has natural angular frequency $\omega_0 = 2\pi\nu_0$ and effective mass m. Classically, this oscillator is radiating with average power $e^2\langle\ddot{\boldsymbol{r}}^2\rangle/6\pi\varepsilon_0 c^3$, where e is the electronic charge and ϵ_0 is the vacuum electric permittivity, and the rate of working of $\boldsymbol{F}$ must be numerically equal to this rate. This enables us to identify $\boldsymbol{F}$ with $e^2\dddot{\boldsymbol{r}}/6\pi\varepsilon_0 c^3$, for with this choice the average rate of working of $\boldsymbol{F}$

$$\langle \boldsymbol{F}\cdot\dot{\boldsymbol{r}}\rangle_{\mathrm{av}} = \frac{e^2}{6\pi\varepsilon_0 c^3}\langle\dot{\boldsymbol{r}}\cdot\dddot{\boldsymbol{r}}\rangle_{\mathrm{av}} = \frac{e^2}{6\pi\varepsilon_0 c^3}\left\langle\frac{\mathrm{d}}{\mathrm{d}t}(\dot{\boldsymbol{r}}\cdot\ddot{\boldsymbol{r}}) - \ddot{\boldsymbol{r}}^2\right\rangle_{\mathrm{av}}$$

and for nearly harmonic motion $\langle\frac{\mathrm{d}}{\mathrm{d}t}(\dot{\boldsymbol{r}}\cdot\ddot{\boldsymbol{r}})\rangle_{\mathrm{av}} = 0$. Hence,

$$\langle \boldsymbol{F}\cdot\dot{\boldsymbol{r}}\rangle_{\mathrm{av}} = -\frac{e^2}{6\pi\varepsilon_0 c^3}\ddot{\boldsymbol{r}}^2 = -(\text{average power radiated}).$$

In a zero-order approximation,

$$\ddot{\boldsymbol{r}} = -\omega_0^2\boldsymbol{r}. \tag{2.2}$$

This enables us to calculate $\boldsymbol{r}$ and so equation (2.1) may be replaced by a simpler one, correct to first order,

$$\ddot{\boldsymbol{r}} = -\omega_0^2\boldsymbol{r} - \gamma\dot{\boldsymbol{r}} \tag{2.3}$$

in which

$$\gamma = \frac{e^2\omega_0^2}{6\pi\varepsilon_0 mc^3} \ll \omega_0$$

that is, the oscillator is lightly damped.

Equation (2.3) is a linear differential equation with constant coefficients, and so we can find a solution in the form $\boldsymbol{r} = \boldsymbol{r}_0\, e^{\alpha t}$. Neglecting γ^2 compared to ω_0^2 we find $\alpha = -\gamma/2 \pm i\omega_0$, so that we may write the solution:

$$\boldsymbol{r} = \boldsymbol{r}_0 \mathrm{e}^{-\gamma t/2} \mathrm{e}^{-\mathrm{i}\omega_0 t}. \tag{2.4}$$

This shows that the oscillator sets up an oscillating but decaying electric field $\boldsymbol{E}(t)$,

$$\boldsymbol{E}(t) = \boldsymbol{E}_0 \mathrm{e}^{-\gamma t/2} \mathrm{e}^{-\mathrm{i}\omega_0 t} \tag{2.5}$$

This is not a pure harmonic oscillation. Introducing $\overline{\boldsymbol{E}}(\omega)$, the Fourier transform of $\boldsymbol{E}(t)$ is

$$\overline{\boldsymbol{E}}(\omega) = \frac{1}{2\pi} \int_{-\infty}^{+\infty} \boldsymbol{E}(t) \mathrm{e}^{\mathrm{i}\omega t} \mathrm{d}t,$$

which shows that a spectrum of frequencies is contained in the radiation. The integral may be evaluated to give

$$\overline{\boldsymbol{E}}(\omega) = \frac{1}{2\pi} \frac{\boldsymbol{E}_0}{\mathrm{i}(\omega - \omega_0) - \gamma/2} \tag{2.6}$$

and the intensity of radiation is

$$I(\omega) = \left|\overline{\boldsymbol{E}}(\omega)\right|^2 \propto \frac{1}{(\omega - \omega_0)^2 + (\gamma/2)^2} \tag{2.7}$$

or, in terms of frequency ν,

$$I(\nu) \propto \frac{1}{(\nu - \nu_0)^2 + (\gamma/4\pi)^2}. \tag{2.8}$$

This form of line shape is clearly peaked at ν_0 and has a width measured by γ. It is called a *Lorentzian* profile.

A quantum mechanical treatment of this problem is to be preferred; it shows that we may identify 2γ with the Einstein A-coefficient which measures the spontaneous transition probability for the relevant transition. This identification is not surprising, since $1/A$ measures the occupation time in the upper level, and by Heisenberg's uncertainty relation, this gives directly the energy width of the line. For strong transitions in the visible, $A \approx 10^8\ \mathrm{s}^{-1}$ so the natural line width is small, $\Delta\nu/\nu \approx 10^{-7}$. Even so, there are situations where this is important, as we shall see in Section 2.2.3.

Doppler broadening. Usually, however, the natural line shape is completely overwhelmed by the well-known Doppler broadening caused by the random velocities of the atoms. The emitted light, at frequency ν_0, suffers a frequency shift to ν due to the components of the atomic velocity v_z in the direction of travel of the photon (say, the z-axis) according to the equation

$$\frac{\nu - \nu_0}{\nu_0} = \frac{v_z}{c}. \tag{2.9}$$

The range of frequencies in the emitted light must correspond to the range of velocities in the gas. If the velocity distribution is Maxwellian, then the number of atoms with z-component of velocities in the range v_z, $v_z + dv_z$ is $dN_z \propto \exp(-Mv_z^2/2kT)dv_z$, where M is the mass of the atom. Consequently, there is a spread in frequency with

$$I(\nu) \propto \exp\left[-(\nu - \nu_0)^2/2\delta^2\right], \tag{2.10}$$

where $\delta^2 = v_0^2 kT/Mc^2$, assuming the radiation travels along the z-axis. Note that the Doppler profile shown in Figure 2.1 is more strongly peaked than a Lorentzian. For the two profiles normalized so that the area underneath

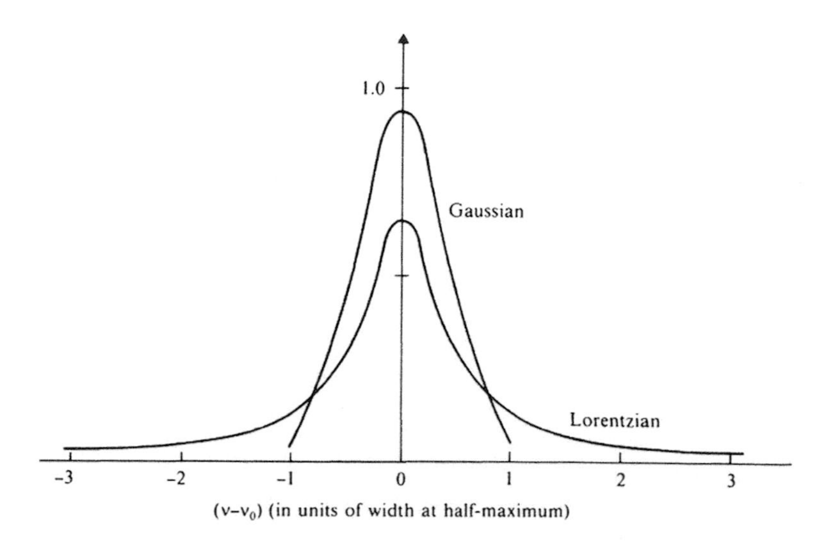

FIGURE 2.1 A comparison of the Doppler (Gaussian) and Lorentzian normalized line shapes. The area under each curve is unity, and each curve has the same width at half maximum, equal to one unit on the frequency axis.

each is unity, at large values of $|\nu - \nu_0|$, the Lorentzian is larger than the Doppler curve. The Doppler profile is also sometimes called a *Gaussian*. This broadening mechanism may contribute in various ways; firstly, as described earlier where the kinetic effects of the atoms produce the broadening; secondly, turbulent effects within a cloud of gas also tend to produce a Gaussian curve of broader shape; finally, if the cloud of gas is bodily in motion relative to the Earth along the line of sight, then all the frequencies are shifted according to equation (2.9) and the broadening occurs around the shifted line centre.

Collisional broadening. This is usually unimportant in the interstellar medium. Classically, we may describe the effect of collisions on the radiating atoms by saying that the radiation train is interrupted, with a loss of phase. The wave trains are now of random length, and so there must be a range of frequencies involved: we need a Fourier sum over frequency to represent the wave, not just a single frequency. Hence, there is a line profile produced by collisional broadening; it is similar to the Lorentzian. The width, however, is $1/\tau_0$, where τ_0 is the mean interval between collisions. Because of the very low density in the interstellar medium, τ_0 may be as much as a thousand years. So this is insignificant when compared with natural broadening in a strong line for which the Einstein A-coefficient may be about 10^8 Hz.

2.2.3 Measurement of Absorption and Emission Lines

The actual profile of an interstellar *absorption* line is usually a combination of a Lorentzian and, possibly, several Gaussians; it may be difficult to measure precisely, particularly for weak lines. What is more easily done is to measure the area in the line, and to calculate the *equivalent width*, W, defined in Figure 2.2: the two shaded areas are arranged to be equal. Mathematically,

$$W = \int \left(1 - \frac{I(\nu)}{I_0}\right) \mathrm{d}\nu \tag{2.11}$$

where I_0 is the intensity of the radiation before absorption. The intensity $I(\nu)$ is controlled by the equation of radiative transfer, which – when emission is negligible – reads for the following plane-parallel case:

$$\mathrm{d}I(\nu)/\mathrm{d}s = -\kappa(\nu)I(\nu). \tag{2.12}$$

$I(\nu)$ is the energy per second in frequency range ν, $\nu + \mathrm{d}\nu$, crossing unit area in unit solid angle; and $\kappa(\nu)$ is a function of frequency; it is the pure absorption coefficient (including scattering) per unit length of path, s. Astronomers measure the amount of absorption suffered in the path, in a particular frequency range, by the optical depth, $\tau_\nu = \int\kappa(\nu)\mathrm{d}s$. Equation (2.12) may be integrated to give

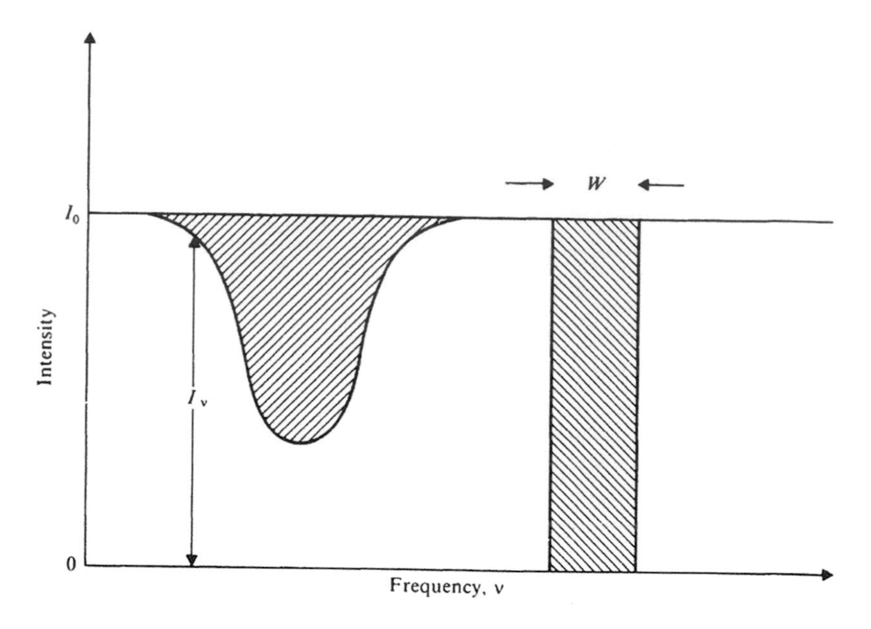

FIGURE 2.2 The equivalent width, W, is chosen so that the two shaded areas are equal.

$$I(\nu) = I_{\nu 0}\mathrm{e}^{-\tau_\nu}, \tag{2.13}$$

where the original intensity before absorption – designated I_0 in Figure 2.2 – may, in reality, vary with frequency, and so it is denoted by $I_{\nu 0}$. Hence, equation (2.11) gives

$$W = \int (1 - e^{-\tau_\nu})\mathrm{d}\nu. \tag{2.14}$$

and if τ_ν is small (the *optically thin* case),

$$W = \int \tau_\nu \mathrm{d}\nu. \tag{2.15}$$

Another way of expressing τ_ν is to introduce the cross-section per atom, σ_ν, which measures the amount of absorption per atom. It has dimensions of area. When we multiply by the total number of absorbers per unit area column, N, we are again measuring the amount of absorption: $\tau_\nu = N\sigma_\nu$. Hence,

$$W = N\int \sigma_\nu d\nu = N\sigma_0\Delta\nu \tag{2.16}$$

where σ_0 is a mean cross-section averaged over the bandwidth $\Delta\nu$. Therefore, a measurement of the area in the line, in the optically thin case, gives a direct method of determining the column density N of the atoms causing the line, if σ_0 is known. The determination of N is usually one of various physical parameters that we seek to obtain from the observational data.

The relationship $W = W(N)$ is called the *curve of growth*, and the portion of it which arises in the optically thin case, equation (2.16), is called the linear portion. When τ_ν is large, as for a strong line, the energy removed from the beam is no longer proportional to τ_ν. If a Doppler profile is inserted for $I(\nu)/I_0$ in equation (2.11), then the dependence of W on N is very weak: $W \propto (\text{constant} + \log N)^{1/2}$. This is called the *flat section of the curve of growth.* If the line is very strong indeed, then absorption in the wings of the line may be important. Here, the profile is the Lorentzian determined by the Einstein A-coefficient. Substitution of the profile in the expression for W yields $W \propto N^{1/2}$, and here we obtain the *square root portion of the curve of growth.* The procedure for astronomers is to make use of the $W(N)$ relation to determine the column densities N. The method depends on the fact that – optically – the absorption is against a point source (a star), but emission is generally into the whole sphere. The fraction of emission directed towards the observer's telescope is, therefore, negligible. Stimulated emission which is concentrated into the same solid angle can be important in certain circumstances (particularly in interstellar masers, see the following text).

For some observations, however, particularly in the radio region, the beam has appreciable angular size and emission from all the atoms in the beam cannot be ignored. The equation of radiative transfer (2.12) then becomes instead

$$\frac{\mathrm{d}I(\nu)}{\mathrm{d}s} = -\kappa(\nu)I(\nu) + j(\nu), \tag{2.17}$$

where $j(\nu)$ is the emissivity, defined so that $j(\nu)\,\mathrm{d}V\,\mathrm{d}\nu\,\mathrm{d}\Omega\,\mathrm{d}t$ is the energy emitted by volume element $\mathrm{d}V$ in the frequency width $\mathrm{d}\nu$ during time interval $\mathrm{d}t$ and into solid angle $\mathrm{d}\Omega$. With $\mathrm{d}\tau_\nu = \kappa(\nu)\mathrm{d}s$, the transfer equation is

$$\frac{\mathrm{d}I(\nu)}{\mathrm{d}\tau_\nu} + I(\nu) = \frac{j(\nu)}{\kappa(\nu)}$$

which is linear and may be integrated by using an integrating factor e^{τ_ν} to give

$$I(\nu) = I_{\nu 0}e^{-\tau_\nu} + \int_0^{\tau_\nu} \frac{j(\nu)}{\kappa(\nu)} \exp\left[-\left(\tau_\nu - \tau_\nu'\right)\right] d\tau_\nu' \tag{2.18}$$

where $I(\nu)$ is defined in Figure 2.3, and $j(\nu)/\kappa(\nu)$ may vary throughout the absorbing region.

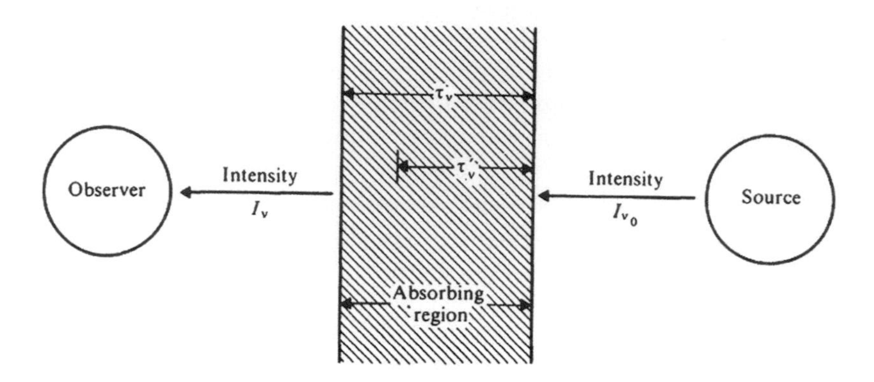

FIGURE 2.3 Schematic diagram defining intensities after absorption.

If the emission is controlled solely by the temperature of the material, as it is when the material is in thermodynamic eqilibrium with its surroundings, then Kirchhoff's law states that

$$j(\nu) = \kappa(\nu)B_\nu(T) \tag{2.19}$$

where $B_\nu(T)$ is the Planck function

$$B_\nu(T) = \frac{2h\nu^3}{c^2}\frac{1}{e^{h\nu/kT}-1} \tag{2.20}$$

that is, the radiation intensity from a black-body at temperature T. Kirchhoff's Law merely says that in thermodynamic equilibrium, emission and absorption are equal. With equations (2.19) and (2.20), equation (2.18) becomes

$$I(\nu) = I_{\nu 0}e^{-\tau_\nu} + B_\nu(T)(1 - e^{-\tau_\nu}). \tag{2.21}$$

Radio astronomers use the concept of *brightness temperature*, T_b, defined as the temperature of the black-body which would give the same intensity $I(\nu)$ in the frequency range observed. At radio wavelengths, $h\nu/kT \ll 1$ and so $e^{h\nu/kT} - 1 \approx h\nu/kT$ so that

$$B_\nu(T) \approx 2\nu^2 kT/c^2$$

(the Rayleigh–Jeans law). Defining T_{b0} be the brightness temperature of the source,

$$T_b = T_{b0}e^{-\tau_\nu} + T(1 - e^{-\tau_\nu}) \tag{2.22}$$

is the brightness temperature measured; T is the temperature defining the populations in the atomic levels involved. If $T_{b0} > T$, then an absorption line appears in the T_b, ν diagram. For, if τ_ν is small at the line centre, then

$$T_b \approx T_{b0}(1 - \tau_\nu) + T\tau_\nu = T_{b0} - (T_{b0} - T)\tau_\nu, \tag{2.23}$$

which shows that T_b decreases as ν approaches the line centre. If τ_ν is large at the line centre, then $e^{-\tau_\nu}$ may be neglected there and so $T_b \approx T$ but it is larger than this outside the line. The area under the line in the T_b, ν diagram is proportional to the number of absorbers in the column.

The *emission* of radiation when τ_ν is small is merely the sum of the emissions per atom. Thus, the total energy emitted into all solid angles per cubic metre per second is $h\nu_{10}n_1A_{10}$ where A_{10} is the Einstein spontaneous transition probability between levels 1 and 0, and n_1 is the number density (m^{-3}) in the excited level 1. The line shape will be that described earlier, and the total energy is the total inside the profile. The problem is in the calculation of n_1. This may be given by a thermodynamic description, so the number in level i at energy ε_i may be

$$n_i = n_0 \exp(-\varepsilon_i/kT), \tag{2.24}$$

where T is the kinetic temperature. Sometimes, however, T is not the kinetic temperature, but a temperature characterizing the mechanisms controlling the level populations; in this case, it is called the *excitation temperature*. Many emission lines from interstellar molecules, in particular, are found to arise in situations which are far from thermodynamic equilibrium. In these cases, the molecules are radiatively or collisionally 'pumped' so that the population n_1 in the excited level involved is considerably greater (e.g. OH) or considerably less (e.g. H_2CO) than the expected value.

How do we know that this is so? Let us consider the example of radio emission from interstellar OH, at a wavelength near 18 cm. If we interpret observations of OH to give the populations of the levels, and assume these populations have arisen as a result of a thermal process, then we may have a rather surprising answer, for the implied temperature from equation (2.24) can be enormous. For OH in some sources in which it is in emission, the inferred temperature may exceed 10^9 K. This is obviously very much greater than can actually be the case. In addition, the radiation is highly polarized and rapidly variable, both phenomena indicative of maser action. To establish a maser in a three-level system between levels 2 and 1 (see Figure 2.4), requires the system to be 'pumped' to some third level, 3, and to cascade into level 2. The intensity is therefore linked to the pumping mechanism, which – if it can be identified – will give information about the radiation intensity at the frequency of the pump, or about the total gas density if the pumping is by collision.

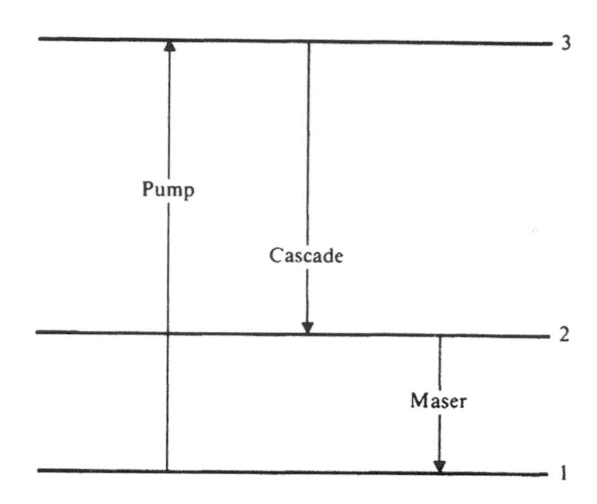

FIGURE 2.4 Schematic diagram illustrating the pumping of a maser emitting between energy levels 1 and 2.

In summary, observations of interstellar lines may give us information concerning some of the following questions:

(i) Is the source cloud in motion relative to the Earth?
(ii) What is the degree of turbulence within the cloud?
(iii) What is the column density of the source atoms or molecules?
(iv) What are the local density and temperature of the source atoms or molecules?

2.2.4 Carriers of Interstellar Lines

Interstellar lines arise from a variety of sources. Atoms and ions usually have lines in the visible and UV regions of the spectrum corresponding to their strongest transitions. For example, the lines of sodium at 589 nm and 589.6 nm, corresponding to its doublet-resonance transition, are seen in absorption in the interstellar medium. The resonance transition in calcium at 422.7 nm is seen in absorption, along with the lines of ionized calcium at 393.4 nm and 396.8 nm. In Figure 2.5, we can see some lines of sodium observed with very high resolution. The different components are believed to arise from sodium atoms in distinct clouds, each with different velocities, along the line of sight from Earth to the star. In the line of sight towards the star ζ Oph, for example, there are more than ten such clouds. In ionized regions of interstellar gas near hot stars, where the temperatures are high and collisions frequent, lines from metastable levels of ionized atoms can be seen.

Lines in the radio spectrum can arise from atoms, ions, and molecules. The most famous radio line is the 21 cm line of atomic hydrogen, which arises in the following way. Each proton and each electron has spin $\frac{1}{2}$. In the H atom

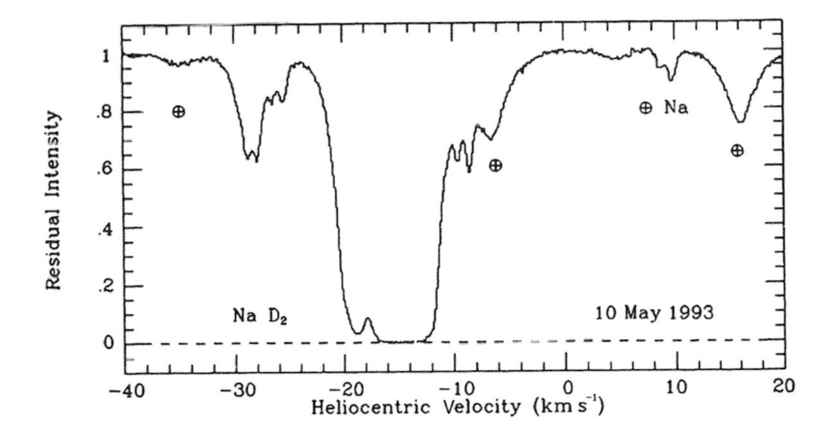

FIGURE 2.5 The interstellar sodium D_2 line observed towards the star ζ Oph. Some telluric lines are indicated by the symbol ⊕. A number of distinct interstellar components are clearly evident. (From Barlow M J et al. 1995 Monthly Notices of the Royal Astronomical Society 272 333.)

ground electronic state, $^2S_{1/2}$, the orbital angular momentum is zero, and so the total spin is either $F = \frac{1}{2} + \frac{1}{2} = 1$ or$F = \frac{1}{2} - \frac{1}{2} = 0$ (the lower energy state). The transition $F = 1$ to $F = 0$ gives rise to the line at wavelength 21.106 cm (frequency 1420.4 MHz), and the Einstein A-coefficient for the transition is 2.8843×10^{15} s^{-1} (= 11.0 Myr)$^{-1}$. In emission, the line is often optically thin, and so its strength is directly proportional to the number of atoms. The natural width is one half of the Einstein A-value, which is quite negligible compared to Doppler broadening. Doppler broadening has a half-width $\Delta\nu \approx 1.01T^{1/2}$ kHz, enabling measurements to define a temperature characteristic of thermal and turbulence behaviour within the cloud. Also, if the cloud is moving relative to Earth in the line of sight with velocity V km s^{-1}, then the central frequency is shifted by an amount $4.738V$ kHz. Since this is usually greater than the broadening, it is easy to resolve separate clouds at different velocities. As discussed in Section 2.2, where absorption occurs, emission must be occurring, too. With a wider beam, the emission can also be detected. Evidently, the same clouds may be detected in both emission and absorption.

Atoms are also important contributors to interstellar radio lines because they can give rise to recombination spectra. When an H atom is ionized, the proton and the electron may recombine, and in so doing cascade through various levels, emitting photons. The so-called Hα line at 656.3 nm, which gives rise to the red colour of many nebulae, is caused by the H^+ + e recombination process when it passes through the stage

$$H(n = 3) \rightarrow H(n = 2) + h\nu,$$

where n is the principal quantum number. Recombination causes transitions between many pairs of levels during the cascade. Since the energy of level n is proportional to $1/n^2$, transitions $n \rightarrow n-1$ will have energies corresponding to the radio region of the spectrum if n is large enough. For example, the transition $n = 109 \rightarrow 108$ is at 5009 MHz. In these states of very high n, near $n = 100$, the electron is normally very far from the nucleus. Recombination through high n levels for atoms other than H atoms occurs in a similar way.

But the main contributors to interstellar lines, especially those which give most information about the denser regions of interstellar space, are the interstellar molecules. Most of the molecular lines observed are lines in the rotational spectrum that are in the radio spectrum. For linear (including diatomic) molecules, the permitted energy levels are approximately those allowed by quantum mechanics for a rigid rotator, that is, a linear molecule that does not change shape as it rotates,

$$E = BJ(J+1), \quad J = 0, 1, 2 \ldots \tag{2.25}$$

with the selection rule

$$\Delta J = \pm 1.$$

The full calculation shows that the constant B is given by

$$B = \frac{h^2}{8\pi^2 I} \tag{2.26}$$

where I is the moment of inertia of the molecule. Thus, massive molecules with large moments of inertia have closely spaced energy levels; for example, cyanodiacetylene, H–C≡C–C≡C–C≡N, has a $J = 4 \rightarrow J = 3$ rotational transition with wavelength 29 mm. Low-mass molecules such as HD, OH, and CH have rotational spectra in the infrared-microwave region ($\lambda \lesssim 1$ mm). For the important molecule CO, the constant B has the value 3.83×10^{-23} J, so that the $J = 1 \rightarrow J = 0$ line has a wavelength of 2.6 mm. This line is observed in many directions in the sky. In fact, after H_2, the CO molecule is the most abundant molecule in the Universe.

Polyatomic molecules are not, in general, linear. If two principal moments of inertia of a molecule are equal (i.e. if the molecule has an axis of symmetry) then it is called a *symmetric top* molecule. In this case, the total energy of rotation consists of rotation about and rotation perpendicular to the symmetry axis. Therefore, two quantum numbers are required: J for the total angular momentum and K for the projection of J on the axis of symmetry. The energy levels in this case are given by

$$E = BJ(J+1) + (A - B)K^2 \tag{2.27}$$

where A, defined similarly to B in equation (2.25), allows for rotation about the figure axis. In equation (2.27)

$$K = -J, -J + 1, -J + 2, \ldots, J$$

and

$$\Delta J = 0, \pm 1 \text{ and } \Delta K = 0.$$

Figure 2.6 shows the schematic level arrangement. It is clearly considerably more complicated than that for the linear molecule.

Asymmetric-top molecules have rotational motion that is even more complicated. Their motion is classified as lying between the extremes of limiting prolate and oblate tops. No simple formulae can be given, and their detailed spectra must be determined by measurement in the laboratory.

Other mechanisms in molecular spectra may give rise to lines that are observed in the interstellar medium. For example, the ammonia molecule, NH_3, in its two lowest vibrational bands provides an example of *hindered vibration*. The plane defined by the three H atoms can be penetrated, quantum mechanically but not classically, by the nitrogen atom in vibration perpendicular to this plane. The resulting double minimum in the potential energy curve gives rise to degenerate rotational levels within these states. The degeneracy is split by the interaction of nuclear and electronic motion, see Figure 2.7. The splitting of rotational levels within the lowest vibrational band gives rise to

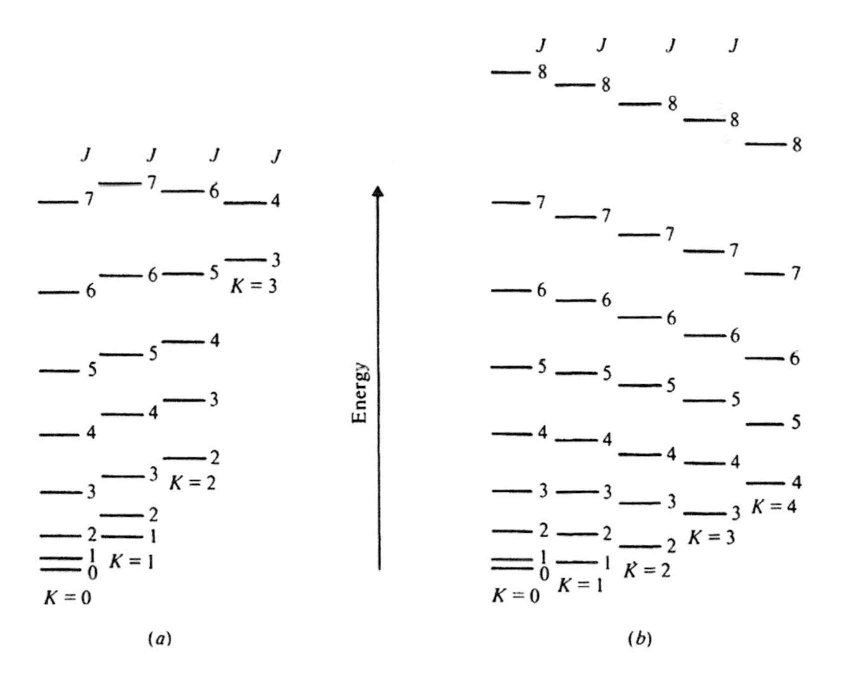

FIGURE 2.6 Rotational energy levels in a symmetric top molecule: (a) A > B (prolate top); (b) A < B (oblate top).

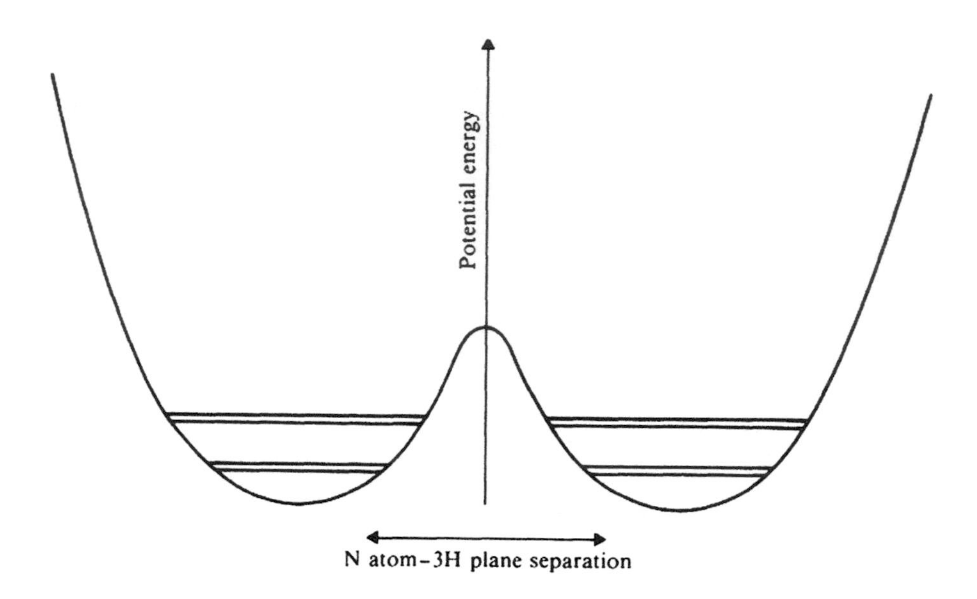

FIGURE 2.7 Schematic potential energy diagram for NH_3. Because of the degeneracy, the levels are split, as shown.

transitions observed in the interstellar medium at a wavelength of approximately 1.25 cm. Rotational transitions of the molecule, however, occur in the infrared. The observation of the radio (1.25 cm) line provides an impressive demonstration of quantum mechanical behaviour. In the OH molecule, also important in the interstellar medium, degeneracy occurs because of the possible choice of ways of adding the electronic angular momentum to that of the rotation of the molecule. The degeneracy is again split by the interaction of these angular momenta, and further hyperfine splitting gives rise to a series of lines near 18 cm in wavelength.

Whatever the mechanism, the process for an astronomer is always the same: to identify the line; to obtain from theory or experiment a value for the absorption cross-section; to calculate the number of atoms, ions, or molecules giving rise to the line; and to infer something about the population of states, and hence, possibly, information about the temperature and density of the regions where these molecules are found.

There are many molecules that are potential probes of warm or cool interstellar gas and although each one of these molecules has many possible rotational transitions, not all of these transitions are equally useful to astronomers. For example, we have noted that carbon monoxide, CO, is widespread in the Galaxy. It is routinely used to trace interstellar clouds with a number density of about 10^9 H_2 molecules m^{-3}. Collisions in the gas between CO and H_2 molecules excite the CO from the ground rotational level, $J = 0$, into the $J = 1$ rotational level, with the energy required coming from the kinetic energy of the hydrogen molecule. In the absence of any other collision, the excited CO will radiate a photon and relax to the $J = 0$ level. The rate of this process is the Einstein A-coefficient for this transition, and the wavelength of the corresponding radiation is, as we said earlier, 2.6 mm.

But what happens if the CO is in gas of much smaller number density? Then collisions are rare and few excitations into the $J = 1$ level occur so the emission at 2.6 mm is weak. On the other hand, what happens in very dense gas? Then collisions occur at a very high rate and CO molecules in the $J = 1$ level may be collisionally de-excited before they can radiate at a rate defined by the Einstein A-coefficient:

$$CO(J = 1) + H_2 \rightarrow CO(J = 0) + H_2 + \Delta E$$

where ΔE is the energy of the transition, now returned to the gas. Evidently, there must be a value of the number density in the gas at which radiation can occur efficiently. This happens when the Einstein A-coefficient for the transition, $A(1\text{–}0)$ s^{-1}, and the rate coefficient of collisional de-excitation, $\gamma(1\text{–}0)$ m^3 s^{-1}, are equal. The number density at which this equality holds is called the *critical density*, n_{crit}, where

$$n_{\text{crit}} = A(1-0)/\gamma(1-0)$$

and A and γ are known quantities. For CO molecules at a temperature of 10 K, n_{crit} for the 1–0 rotational transition is 1.8×10^9 m^{-3}, so that CO observations at 2.6 mm preferentially identify cold clouds with number densities around this value.

Each molecular transition used as a probe of the interstellar gas at a particular temperature has its own critical density. Thus, specific molecules can be used to identify gas in a particular range of number density and temperature. For example, the critical density for CO(6–5) at 100 K is 2.5×10^{11} m^{-3}, for carbonyl sulfide, OCS(8–7) at 20 K is 3.5×10^{10} m^{-3}, for formyl cation, HCO^+(2–1) at 10 K is 1.1×10^{12} m^{-3} and HCO^+(7–6) at 100 K is 4.9×10^{13} m^{-3}, and cyanoacetylene, HC_3N (12–11) at 20 K is 7.1×10^{11} m^{-3} and HC_3N(19–18) at 80 K is 2.9×10^{12} m^{-3}.

2.3 CONTINUUM RADIATION

There are various kinds of continuum radiation that arise in the interstellar medium, and continuum emission and continuum absorption can both occur. Ionization gives rise to continuous absorption by the bound-free and free-free processes which may occur in ions, atoms, and molecules, shown schematically in Figure 2.8. Ionization continua are important in the spectra of stellar atmospheres and planetary nebulae, but less important in the interstellar medium. Dissociation continua of molecules control in an important way the penetration of UV radiation into the interior of an interstellar cloud containing H_2 molecules (see Section 3.3.2). But neither of these processes is important in defining the constitution of the interstellar medium in the same way as

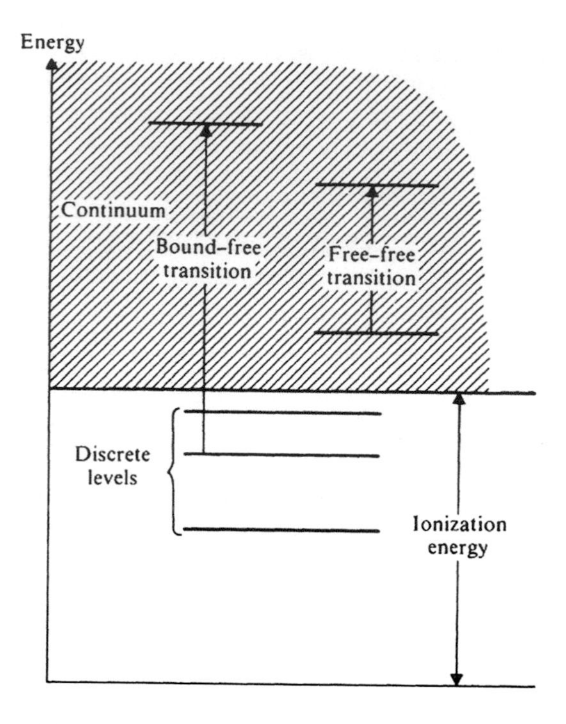

FIGURE 2.8 Bound–free and free–free transitions illustrated in a schematic energy level diagram.

line processes do. They are, however, important in the description of ionized regions near hot stars (see Chapter 5).

Bremsstrahlung and synchrotron radiation are other continuous emissions from different origins. Bremsstrahlung is radiation emitted by charged particles when they are accelerated during collisions. It arises in the interaction of very fast charged particles with a gas, and also is emitted from a plasma. The associated energy loss by radiation depends on the energy in the colliding pair and some other parameters. For a Maxwellian energy distribution in a plasma at temperature T, the emission in the frequency range ν, $\nu + d\nu$ is proportional to $T^{-1/2} \exp(-h\nu/kT)n_e n_i$, where n_e and n_i are the number densities of electrons and ions. Thus, the main contribution occurs when the photon energy $h\nu \leq kT$. The process may also act in absorption and must be taken into account when observing radio sources. We discuss bremsstrahlung later (Section 5.4.1), in connection with ionized nebulae.

Synchrotron emission is caused when a relativistic electron is gyrating about a magnetic field line. An electron of charge e projected with velocity $\boldsymbol{V}$ perpendicular to a magnetic field line experiences a Lorentz force proportional to $e\boldsymbol{V} \times \boldsymbol{B}$ where $\boldsymbol{B}$ measures the magnetic field. The Lorentz force is at right angles to the direction of travel and so causes circular or helical motion if $|\mathbf{V}|$ and $\boldsymbol{B}$ are constant. Because of the acceleration suffered, the electron will radiate, and the radiation is emitted in a highly forward direction. The cone containing the radiation sweeps past the observer very rapidly and this modifies the frequency spectrum. The importance of this radiation for studies of the interstellar medium is that it gives direct evidence for a magnetic field and for a flux of cosmic ray electrons at very high energies. These electrons, and more particularly the corresponding cosmic ray protons, have a profound effect on the processes within interstellar clouds (see Chapter 3).

In addition, the Universe is bathed in another form of continuous radiation; this is found to be electromagnetic radiation typical of a 2.7 K black-body. It is believed to be the relic radiation from the era of recombination in the Early Universe. The peak intensity occurs at a wavelength of approximately 1 mm. The energy content per unit volume in this radiation is substantial, and is comparable to that in stellar radiation in the Galaxy, to that in gas motions, in cosmic rays, and also in the magnetic field. However, the coupling between interstellar gas and the black-body radiation is usually weak. Nevertheless, for radiative processes occurring at $\lambda \approx 1$ mm, this is an important radiation source. For example, rotational transitions in CN radicals are affected by this background. Formaldehyde can be seen in absorption against this background because of collisions which tend to depopulate the upper level of the transition.

2.4 INTERSTELLAR EXTINCTION

Optical photographs of rich star fields in our Milky Way Galaxy show dark regions where it seems that stars do not shine (Figure 2.9, left-hand image). These regions could be places where there are no stars, or, alternatively, they could be places where stars are present but are obscured from our view. The latter explanation is now accepted, for closer examination of many dark regions does show that stars are present but that their light is heavily extinguished. Also, observations made at longer wavelengths rather than optical (i.e. in the infrared wavelengths) reveal the presence of stars and confirm that the phenomenon causing extinction varies with wavelength (see Figure 2.9, right-hand image), being weaker at longer wavelengths. It also appears that extinction is not localized within these dark regions, but is widespread throughout the Galaxy.

In Chapter 4, we shall discuss this and other evidence in more detail. We shall see that it points to the existence of yet another component in the interstellar medium: *interstellar dust grains.* These are small solid particles, of radius from about 10^{-9} to 10^{-7} m, containing in all about 1% of the interstellar mass. These grains are believed to be well mixed with the gas. Where the gas is denser, then those regions also have more dust grains, and the extinction is greater. Figure 2.9 shows a region where the gas is denser and consequently starlight is more heavily extinguished.

It will become apparent that grains, for all their tiny mass and low spatial density (on average, about one dust grain per 10^6 m^3 – exceptionally clean, by terrestrial standards), are an important component of the interstellar medium, not only with regard to their effects on radiation. We shall see that they also play an important role in the evolution of the interstellar gas (see Chapter 4).

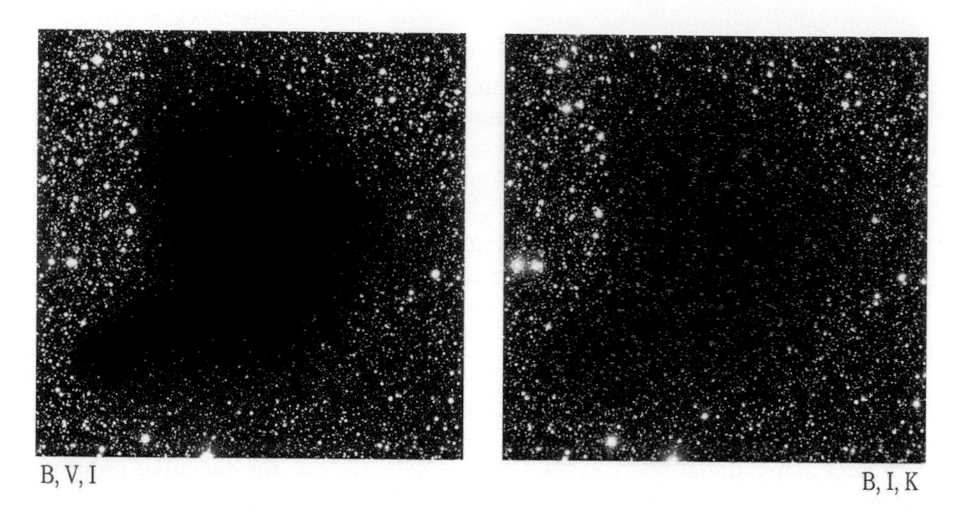

FIGURE 2.9 A dark cloud, Barnard 68, in optical (left-hand image) and infrared (right-hand image) radiation. (Credit for both images: ESO.)

2.5 SOME INTERSTELLAR REGIONS IDENTIFIED: INTERSTELLAR CLOUDS

What is the picture of the interstellar medium that emerges from these various observations? We know that the medium is filled with a gas of mean density about 10^6 atoms m^{-3} and that it is certainly not uniform, but very clumpy. The gas is mainly hydrogen; about 10% of the atoms are helium; and a further 0.1% of atoms are carbon, nitrogen, or oxygen. Other elements are even less abundant. Mixed with the gas are the dust grains. The ratio (by number) of gas atoms to dust grains of about 10^{12} seems fairly uniform for the regions where measurements can be made.

The range of densities and temperatures found in the interstellar medium is very large. Some extended regions have very low densities, about 10^3 atoms m^{-3}, and very high temperatures, about 10^6 K, these very high temperatures being deduced from observations of lines of multiply ionized atoms such as interstellar oxygen, O^{5+}. Such high temperature regions are, of course, completely ionized. Other hot regions are somewhat cooler ($T \sim 10^4$ K) and denser ($\sim 10^5$ atoms m^{-3}) and partially ionized. They can be detected using the H atom 21 cm line. Gas close to and heated by hot stars radiate strongly. These near-stellar regions are called diffuse nebulae or HII regions, this spectroscopic symbol denoting that the gas is mainly ionized hydrogen, H^+.

Cooler, denser matter is contained in the mainly neutral regions also observed by the H atom line and lines of other atoms and ions; these clouds have densities of 10^7–10^9 atoms m^{-3} and temperatures of about 100 K. They may be some parsecs in size, but no 'standard' cloud exists; the cloud parameters vary significantly. The interiors of these clouds are partially shielded from starlight by extinction because of the dust grains they contain. They also contain some simple molecules, such as H_2 and CO.

Interstellar molecules are generally found in regions which seem to be denser than these clouds. These regions are often deficient in H atom 21 cm radiation, so it is inferred that most of the hydrogen atoms have been combined into hydrogen molecules. These denser clouds are dark; little starlight penetrates to the interior. The densities in the dark clouds apear to be 10^9–10^{10} molecules m^{-3} and the temperatures are quite low, about 10 K. These clouds may associate to form some of the most massive objects known in the Galaxy, the so-called Giant Molecular Clouds, with masses up to about a million solar masses.

Still denser regions are observed, where a huge variety of even more complex molecules are found. These dense cores seem to be associated with regions where massive stars are forming. Temperatures in these dense cores are higher (at about a few hundred K) than in the dark clouds and densities may be in excess of 10^{13} molecules m^{-3}. Figure 2.10 shows an image of an interstellar cloud in the constellation of Cepheus in which star formation is occurring.

FIGURE 2.10 The image shows an interstellar cloud in the constellation of Cepheus. In the optical, this region appears very black because of interstellar extinction, but in the infrared as shown the cloud is bright because of emission from warm dust grains. The infrared observations were from NASA's Spitzer Space Telescope. X-ray observations (from the Chandra X-ray Observatory) show (in violet) young hot stars in the region. (Credits: NASA/CXC/PSU/K Geltman et al.; NASA/JPL-Caltech/CfA/J Wang et al.)

Tables 2.1(a) and (b) summarize information about the main hot and cold components of gas in the interstellar medium. It is important to remember that the observed properties of these interstellar components cover a wide range and the values quoted in the tables are approximate and intended to be merely indicative. We note in passing that the atmospheres of cool stars and their winds are also largely molecular and can be dusty, and – rather surprisingly – ejecta of supernovae and novae can also be dusty and molecular. However, we shall not be concerned with those topics in this book.

The main electromagnetic fields with which we shall be concerned are those of starlight. Hot stars, although they are relatively few in number, dominate the interstellar radiation field in the ultraviolet into the visible. More numerous cooler stars radiate in the visible and infrared. The very hot ionized regions generate X-rays, relativistic electrons generate synchrotron emission, and the diffuse nebulae have powerful optical and ultraviolet emissions. Dust grains heated by starlight generate emission in the infrared and far infrared. The whole Universe is bathed in the cosmic microwave background radiation, a black-body spectrum with a temperature of 2.7 K.

The whole of the interstellar medium is penetrated by cosmic rays which contain mainly electrons and protons of very high energy. These particles each have

TABLE 2.1 (a)

Indicative values of physical parameters of the main hot components of interstellar gas

Name	Temperature (K)	Number density (m^{-3})	State of hydrogen	Typical dimension (pc)	Filling factor (%)
Coronal gas	$\sim 10^6$	$\sim 10^3$	H^+	1000	~50
Warm ionized gas	$\sim 10^4$	$\sim 3 \times 10^5$	H^+	1000	~40
Warm neutral gas	$\sim 10^4$	$\sim 3 \times 10^5$	H	100	~10
HII regions around hot stars	$\sim 10^4$	$\sim 10^8$–10^{10}	H^+	1	<1

TABLE 2.1 (b)

Indicative values of physical parameters of the main cold components of interstellar gas

Name	Temperature (K)	Number density (m^{-3})	Main state of hydrogen	Typical dimension (pc)
Cool neutral gas	~100	$\sim 3 \times 10^7$	H	~10
Diffuse cloud	~100	$\sim 10^8$	H, H_2	~10
Molecular cloud	~10	$\sim 10^{10}$	H_2	~0.1
Star-forming gas	~100–300	$\sim 10^{13}$	H_2	~0.01

remarkably high energies, in extreme cases up to about 10^{20} eV. However, the flux declines steeply with energy and lower energy cosmic rays (at about a few MeV) are the most important in causing ionization in interstellar clouds.

The Galaxy carries with it a magnetic field that is somewhat chaotic but with some regular structure. The magnetic field is about 10^{-9} T in the solar neighbourhood, but may be as large as $\sim 10^{-6}$ T in very dense clouds in star-forming regions. Magnetic fields may have important consequences in interstellar gas dynamics and in star-forming regions.

While the total mass of the Milky Way galaxy is about 10^{12} $M_\odot$, the total mass of the interstellar medium is about 10^{11} $M_\odot$. Most of the interstellar mass is gas; the mass of dust in the interstellar medium is about 10^9 $M_\odot$. Other galaxies have a wide range of masses, and may be proportionally richer or poorer in interstellar gas and interstellar dust than the Milky Way.

The sequence of interstellar densities listed in Tables 2.1(a) and (b) traces a potential evolutionary process in which tenuous gas is compressed to form stars, some of which ultimately explode as supernovae, heating and dissipating the gas. It suggests that, although the material in the interstellar medium seems insignificant compared to that in stars, it is extremely important in galactic evolution. We shall return to this problem in Chapter 8. Once a massive star is formed in the interstellar medium, it has dominating effects on the gas surrounding it. In particular (Chapter 5), the gas is ionized by the star's radiation, and the effect of this on the interstellar medium is to produce some major dynamical effects. We study these processes in Chapter 7.

Problems

1. Show that the full width of the line profile at half-maximum height (FWHM) is given by $\gamma/2\pi$ for the Lorentzian (equation (2.8)) and by 2.3555δ for the Gaussian (equation (2.10)).
2. What fraction of total energy is radiated in an emission line of Lorentzian profile within a bandwidth defined by the natural linewidth centred on the line centre?
3. With the definition of equivalent width, W, as in equation (2.11), and assuming that the optical depth at frequency ν is $\tau_\nu = \tau_0/(\nu - \nu_0)^2$, show by a suitable transformation that $W \propto \tau_0^{1/2}$.
4. Calculate the Doppler FWHM for a gas of H atoms radiating at a wavelength of 100 nm at a temperature of 300 K. Show that collisional broadening in such a gas will not become comparable until the number density is approximately 10^{27} m^{-3} (use a geometric cross-section for H atom collisions).
5. Derive equations (2.13), (2.15), (2.18), and (2.22) from the preceding equations.
6. Find the wavelength (in mm) of the $J = 1 \rightarrow J = 0$ rotational transition in $^{12}C^{32}S$, given that the equilibrium C–S separation is 1.535×10^{-10} m.

3 Microscopic Processes in the Interstellar Medium

3.1 INTRODUCTION

We have seen in Chapter 2 how microscopic processes give information to astronomers about the bulk properties of the interstellar medium; for example, observations of the 2.6 mm line of the CO molecule may tell us about the temperatures and densities of the regions in which the line is formed. However, these microscopic processes do more than that: they also *control* the bulk properties of the medium, such as temperature and density. In determining these properties, they also determine the *evolution* of the interstellar material and so – to a large extent – they determine, for example, the sites of star formation and other large-scale features of the Galaxy. In this chapter, we discuss microscopic processes, especially those which directly or indirectly determine the temperature of the interstellar gas, for temperature is a dominating parameter.

3.2 COOLING OF THE INTERSTELLAR GAS

3.2.1 How Do Gas Clouds Cool?

Interstellar clouds cool by emitting radiation; they do not usually conduct or convect heat very efficiently. The mechanism by which this radiation occurs is usually initiated by an excitation of an atomic, ionic, or molecular transition during a collision. In this excitation, the atom, ion, or molecule gains its energy from the kinetic energy of the colliding pair. After a time, the excited system radiates this energy away in a photon, which may escape from the cloud. Thus, the gas loses kinetic energy, and so it cools. We summarize the process by the equations

$$\mathrm{A} + \mathrm{B} \rightarrow \mathrm{A} + \mathrm{B}^* \tag{3.1}$$

$$\mathrm{B}^* \rightarrow \mathrm{B} + h\nu. \tag{3.2}$$

We look briefly at the physics of this process, to see what kind of transitions are likely to be important, and what pairs of colliding partners lead to efficient cooling. It is clear that the most efficient cooling processes are likely to be those in which the following criteria are satisfied:

(i) frequent collisions; implying fairly abundant partners,
(ii) excitation energy comparable to or less than the thermal kinetic energy,
(iii) a high probability of excitation during the collision,
(iv) the photon is normally emitted before a second collision occurs on the excited partner, and
(v) the photons emitted are not re-absorbed; that is, the gas is said to be optically thin in the cooling radiation.

3.2.2 Cooling by Ions and Atoms

Criterion (i) tells us that important collisions are likely to be those in which hydrogen and the more abundant atoms (C, N, O) or their ions, and electrons are involved, under appropriate circumstances. Criterion (ii) tells us that, for example, where the kinetic temperature is about 100 K (as it is in many low-density clouds), then the excitation energy should be the equivalent of about 100 K also, for efficient cooling. The predominant form of carbon in many regions is C^+, which has a transition $^2P_{1/2} \rightarrow {}^2P_{3/2}$ with energy difference $\Delta E = 1.4 \times 10^{-21}$ J $\equiv$ 92 K. Clearly, the excitation of this transition may be important in clouds with temperatures around 100 K, though not in clouds with temperatures of, say, 20 K, in which most collisions would be insufficiently energetic to cause transitions in C^+. If the energy criterion (ii) is satisfied, then we might expect that collisions of electrons with C^+ would be important because of the large Coulomb forces between these partners. To examine criterion (iii) requires a proper quantum mechanical calculation of the excitation cross-section. However, the cross-section is usually large when the transition is allowed, and this is the test for criterion (iv) also. Where the transition is optically allowed (has a non-zero electric dipole moment), we can think of the electric field of the passing electron as causing the transition. Where there is zero dipole moment, the electron may cause a spin-change in the ion or atom. Criterion (iv) is not always satisfied, however, as we shall see in the case of H_2 cooling. We show in Table 3.1 several of the transitions important in clouds with temperatures near 100 K.

TABLE 3.1

Some important cooling transitions in cool interstellar clouds ($T \simeq 100$ K)

Transition	Colliding partners	$\Delta E/k$
C^+ ($^2P_{1/2} \rightarrow {}^2P_{3/2}$)	H, e, H_2	92 K
Si^+ ($^2P_{1/2} \rightarrow {}^2P_{3/2}$)	e	413 K
O ($^3P_2 \rightarrow {}^3P_{1,0}$)	H, e	{228K 326K

To calculate the cooling rate, we need to know the collisional excitation cross-section as a function of temperature. Assuming that criterion (v) is satisfied, then in the case of e and C^+, the cooling rate turns out to be about

$$\Lambda_{C^+} = n(e)n(C^+)8 \times 10^{-33} T^{-1/2} \exp(-92K/T)\, \mathrm{J\,m^{-3}s^{-1}} \qquad (3.3)$$

The exponential arises because of the Maxwellian distribution of velocities.

Since hydrogen is the most abundant element, excitation of transitions in atomic hydrogen – if it occurs – is likely to be an important cooling mechanism. However, the transitions are so energetic (more than 10 eV above the ground state) that only at high temperatures (near 10^4 K) does this mechanism begin to play a role. At intermediate temperatures (near 10^3 K), it is found that ions such as Fe^+ and Si^+ may be excited by collisions into long-lived metastable states which act as new ground states from which excitation occurs.

3.2.3 Cooling by Molecules

Molecular hydrogen can be very abundant in interstellar gas, and may therefore also be an important coolant. It possesses a spectrum of lines due to the rotation of the molecule. As we saw in Chapter 2, a rigid rotator will have energies restricted to a discrete set of values

$$E_J = BJ(J+1),\ J = 0, 1, 2\ldots \qquad (3.4)$$

For H_2, electric dipole transitions between these levels are forbidden, because no dipole moment exists in the molecule in these states. Transitions occur via electric quadrupole interaction, and the selection rule for these transitions is $\Delta J = \pm 2$. The least energetic transition in H_2, the excitation $J = 0$ to $J = 2$, occurs at an energy equivalent to 510 K. The cooling process by H_2 is, however, different from the ion cooling processes described earlier. This is because the lifetimes of these rotational levels against radiation by quadrupole interaction are so long – for example, 3×10^{10} s for the $J = 2$ level – that at typical densities the molecules are populated due to collisions which occur at intervals of typically about $10^{11}/n$ s, which is normally shorter than the radiative relaxation rate. The levels therefore become populated with a population, N_J, where

$$N_J \propto (2J+1)\exp(-E_J/kT) \qquad (3.5)$$

characteristic of the kinetic temperature. The radiation leaks out slowly from this distribution and is only a minor perturbation to the level populations. This radiation is very unlikely to be re-absorbed by H_2 elsewhere in the cloud.

These arguments, which apply to H_2, do not apply to HD, the deuterium-substituted molecule. HD does have a small dipole moment because its centre of mass is not at its centre of charge. This means that transitions between its rotational levels are allowed, with $\Delta J = \pm 1$, and HD is more effective at cooling, per molecule. However, the HD abundance is usually low and it will not add significantly to the cooling rate.

The cooling rate for pure H_2 is

$$\Lambda_{H_2} = \sum_{J \geq 2} n(H_2, J)\Delta E(J \rightarrow J - 2)A(J \rightarrow J - 2) \tag{3.6}$$

which, at 100 K, gives a value of about 3×10^{-33} J s^{-1} per H_2 molecule.

Other molecules are also very important coolants, especially in dark clouds. We know that, after H_2, the next most abundant interstellar molecule is CO. In fact, in dark dense clouds in which all hydrogen is molecular and $n(H_2) \gtrsim 10^{10}$ m^{-3}, then, typically, $n(CO) \approx 10^{-5}n(H_2)$. Considerable amounts of CO also exist in lower-density clouds. CO is the most important coolant in denser clouds, even at a fractional abundance relative to hydrogen of only $\sim 10^{-5}$, because CO possesses a dipole moment and so its rotational transitions are permitted. The CO molecule therefore relaxes to its ground state very quickly, cooling the gas and being then available for another excitation. The lowest CO rotational energy levels, $J = 0$ and $J = 1$, are separated by an energy equivalent to 5.5 K, so that CO should be an effective coolant down to temperatures of this order. However, the column density of CO in dense clouds is high, and CO may then become an efficient absorber of its own photons, and criterion (v) is not satisfied. When such radiation trapping occurs (in which case the radiation is said to be *optically thick*), the efficiency of CO cooling is much reduced and other less abundant molecules (e.g. OH and H_2O, or minor isotopes of CO such as ^{13}CO or $C^{18}O$) may contribute significantly to the cooling.

Note that other cooling mechanisms dominate at temperatures of about 10^4 K if the gas is almost fully ionized. Such processes are discussed in Chapter 5.

3.3 HEATING OF THE INTERSTELLAR GAS

3.3.1 How Are Interstellar Gases Heated?

There are several sources of energy for the interstellar gas, including starlight (either localized near a star where the radiation field is intense, or a mean intensity throughout the Galaxy), cosmic rays, X-rays, transient events such as stellar winds, kinetic energy from explosions of novae and supernovae, and compressional and frictional heating.

3.3.2 Heating by Starlight

Every process of photoionization by radiation of frequency ν of a system with ionization potential I, where $h\nu > I$

$$A + h\nu \rightarrow A^+ + e \tag{3.7}$$

yields an electron e with energy $(h\nu - I)$. This electron interacts with the gas: if most of its interactions are elastic collisions with atoms and ambient electrons, then the energy ($h\nu - I$) becomes shared with the gas and constitutes a heat source. Some photoelectrons, however, excite transitions in ions or atoms which subsequently re-radiate, and such energy will be lost. Despite this, there is an overall contribution to the heating of cool clouds by the photoionization of atoms such as C, Si, and Fe. In HII regions (in which the H atoms are almost completely ionized), the most important contribution is made by the ionization of H atoms, and we shall discuss these regions in Chapter 5. We shall see that these HII regions capture all photons with energy larger than 13.6 eV. In HI regions (in which the H atoms are mainly neutral), therefore, photons always have energy less than 13.6 eV (the ionization potential of hydrogen atoms). Thus, in HI regions, each ionization of carbon (ionization potential 11.3 eV) produces an electron of energy at most 2.3 eV. The mean value, for a gas of cosmic abundance, of heat deposited per ionization is about 2.1 eV.

A slightly different type of heating can occur when H_2 is photodissociated. The mechanism is a state-to-state excitation from the ground state X in the lowest vibrational level to an excited state B, followed by a cascade into the vibrational continuum of the ground state X, in which the molecule dissociates (see Figure 3.1). A significant fraction (~ 20%) of all such excitations

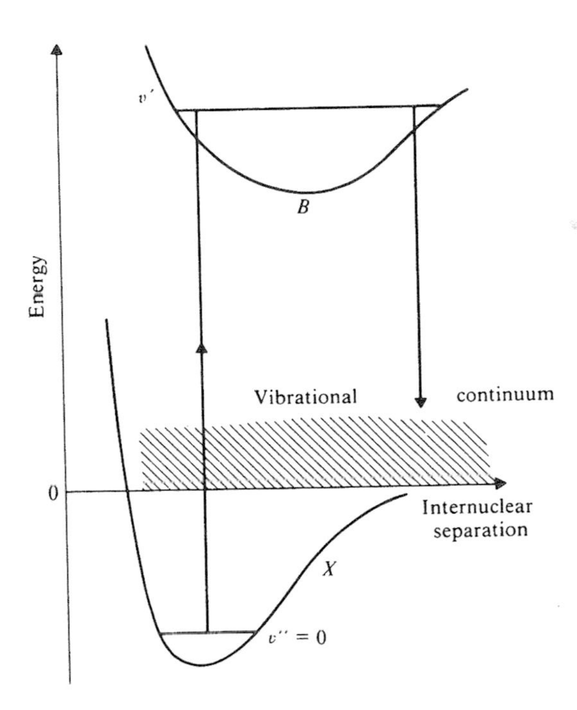

FIGURE 3.1 A schematic representation of the potential energy curves of the H_2 molecule ground electronic state X and the first excited bound electronic state B. These states are involved in the photodissociation of H_2 in interstellar clouds. The molecule in the vibrational continuum 'falls apart', with about 0.4 eV of energy in the dissociating H atoms. Most of the transitions from state B fall into bound excited vibrational levels of the groundstate, X.

leads to dissociation. The dissociating atoms have kinetic energy which then becomes distributed in the gas. The mean energy released per dissociation can be calculated to be about 0.4 eV, and this provides a heat source in regions of space where hydrogen molecules are formed and destroyed rapidly.

3.3.3 Heating by Cosmic Rays and X-Rays

Cosmic ray particles consist primarily of high-energy protons and electrons, the most abundant of those that we can detect on Earth having energies of a few MeV. Soft X-rays occur with a range of photon energies, but their greatest intensity is around 0.1 keV.

Cosmic ray protons and X-rays can ionize H atoms:

$$(\mathrm{p}, \mathrm{X}) + \mathrm{H} \rightarrow (\mathrm{p}', \mathrm{X}') + \mathrm{H}^+ + \mathrm{e}. \tag{3.8}$$

For a 2 MeV proton, the electrons arising in reaction (3.8) have a wide distribution in energy, with a mean energy of about 30 eV. However, not all of this energy will appear as heat unless the medium is mostly ionized, when the energy is mainly dissipated in elastic collisions. In a mainly neutral medium, however, the electron may cause excitations or (if the electron is sufficiently energetic) further ionization, for example,

$$\mathrm{e} + \mathrm{H(1s)} \rightarrow \mathrm{e} + \mathrm{H(2p)} \rightarrow \mathrm{e} + \mathrm{H(1s)} + h\nu \tag{3.9}$$

or

$$\mathrm{e} + \mathrm{H(1s)} \rightarrow \mathrm{e} + \mathrm{H}^+ + \mathrm{e} \tag{3.10}$$

In the excitation (3.9), the photon emitted will generally escape from the gas, and that energy is lost. Elastic collisions between the energetic electrons and the ambient electrons will slowly share some of the energy with the gas. When the electron energy has been reduced below 13.6 eV, ionization can no longer occur, and when below 10.2 eV, excitation of H atoms is no longer possible. When allowance is made for the energy-loss processes and the production of secondary electrons, it is found that about 3.4 eV of kinetic energy is injected per electron (primary or secondary) produced.

The situation for X-ray ionization is similar, except that helium, which is present at about 10% by number of H atoms, now plays a significant role because the cross-section for absorption of X-rays by He atoms is substantially larger than for H atoms. The electron released in the ionization can share its energy with the ambient electrons or produce further ionization and excitation. Detailed calculation shows that although a 50 eV X-ray photon colliding with a He atom releases a 25 eV electron, in fact only about 6 eV of that energy is deposited as heat in the limiting case of zero fractional ionization.

The physical processes involved in the ionization of a molecular hydrogen cloud are much more varied. Cosmic ray or X-ray ionization of H_2 (which has an ionization potential of 15.4 eV) produces H_2^+, which rapidly undergoes reaction with H_2 to form H_3^+, the protonated molecular hydrogen ion:

$$H_2^+ + H_2 \rightarrow H_3^+ + H \tag{3.11}$$

which, as we shall see later, plays a very important role in the chemistry of interstellar dark clouds. This ion can be destroyed rapidly in dissociative recombination with electrons:

$$H_3^+ + e \rightarrow 3H \text{ or } H_2 + H. \tag{3.12}$$

Both of these reactions are exothermic (otherwise they would not proceed in interstellar conditions) and the total excess of energy in the products is about 11 eV. Some of this energy may be locked up in internal modes of the molecules, but about two-thirds of it is available for heating. This is in addition to some fraction of the energy of the electron released in the ionization of H_2. This electron can also partake in a variety of reactions with H_2, for example,

elastic scattering:

$$e + H_2 \rightarrow e + H_2 \tag{3.13}$$

inelastic scattering:

$$e + H_2 \rightarrow e + H_2^* \tag{3.14}$$

(exciting electronic, vibrational, or rotational transitions in H_2),

ionization:

$$e + H_2 \rightarrow e + H_2^+ + e \tag{3.15}$$

(where an additional electron is released and the H_2^+ produced may be internally excited),

and dissociation:

$$e + H_2 \rightarrow e + H + H \tag{3.16}$$

(where the H atoms may be electronically excited in this process).

Relaxation of electronically excited H_2 produced in reaction (3.14) gives rise to an ultraviolet radiation field that can be important deep inside dark clouds where starlight cannot penetrate, though cosmic ray protons do. This radiation (called the *cosmic ray induced radiation field*) is of low intensity, about 10^{-4} of the mean interstellar radiation field intensity in the Milky Way galaxy. Nevertheless, it can cause significant amounts of dissociation of interstellar molecules in the interiors of dark clouds from which starlight is excluded.

Secondary electrons released in reaction (3.15) may also be able to take part in some or all of these reactions. Since all the possible routes are energy dependent, the various cross-sections and their energy dependence must be known to calculate the mean overall effect per ionization. When this is done, it is found that a 2 MeV proton produces about 15 eV of heating after an ionization of H_2, in which event the proton loses 35 eV. For a fractional ionization of 10^{-3}, this rises to 17 eV, owing to the additional thermalizing elastic collisions between electrons. For X-rays, calculations show that about 35% of their energy is a heat source for the gas. We see that the presence of H_2 ensures that considerably higher heating rates occur than in a pure H atom cloud. This compensates to some extent for the more powerful cooling mechanisms in such clouds, though denser clouds are generally found to be cooler than more tenuous ones.

3.3.4 Grains as a Heating Mechanism?

We shall discuss the properties of interstellar dust grains in Chapter 4. But we note here that if photoelectric emission occurs at the surface of interstellar grains then each event releases to the gas an electron with a few electron volts of energy, equal to the difference in energy between the photon and the work function of the grain material. Depending on the photoelectric efficiency, the process can be one of the important heat sources in diffuse, mainly neutral clouds (see Section 4.5).

3.3.5 Temperatures in Interstellar Regions

The characteristic temperatures for some interstellar regions that were shown in Table 2.1 are reasonably consistent with temperatures computed from what is known of the heating and cooling mechanisms operating in those regions.

3.4 MOLECULE FORMATION

3.4.1 The Formation of Interstellar Molecules in Gas Phase Reactions

Since the formation and the presence of molecules are important for the properties of interstellar clouds, we shall now consider the reactions by which they arise through purely gas phase reactions. Surface reactions on dust grains will be considered in Chapter 4. At the temperatures, densities, and radiation intensities in the interstellar medium, molecules do not form as readily as one might expect, and the product molecules do not persist indefinitely.

Consider the interaction of two atoms, A and B. If they collide, then they may interact along a potential energy curve which is attractive (see Figure 3.2). If no energy is removed from the colliding pair then, of course, they will merely bounce apart again, and no molecule will be formed. The time in contact is very short, comparable to the period of a vibrational oscillation in the molecule, typically of the order of 10^{-13} s. The complex does not easily radiate energy away on this timescale, even supposing a suitable transition exists (such as from the solid to the dashed curve in the figure). The largest transition probabilities are usually about 10^8 s^{-1}, so the greatest probability of radiative stabilization, where it is possible, is $\lesssim 10^{-5}$ per collision; that is, at best only one in 100 000 collisions of A and B produces the molecule AB.

Another way to stabilize the colliding pair as a molecule is to require a third body to remove some energy during the collision. This is the normal situation in reactions at atmospheric densities because a third body will almost certainly impact on a colliding pair before they have separated. However, under almost

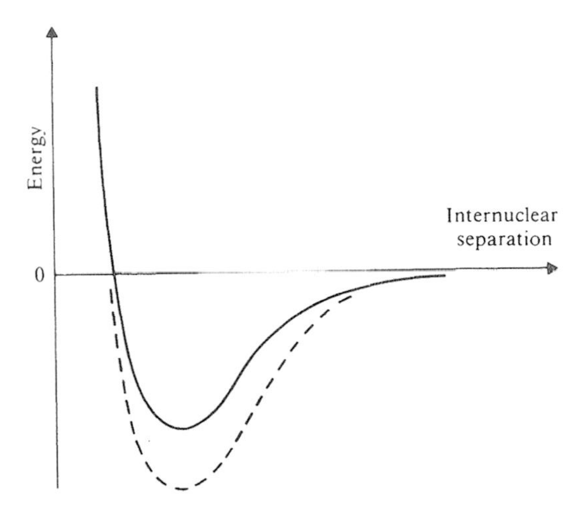

FIGURE 3.2 A schematic representation of the potential energy interaction of two atoms, as a function of internuclear separation.

all interstellar conditions, this is a forlorn hope; one can show that the number of collisions on the colliding pair AB during the time they are in contact is approximately $10^{-29}n$, where n in the number density per m^3 of the most abundant collision partner (usually H_2): therefore, three-body collisions of gaseous atoms are usually negligibly rare in the interstellar medium, except in cool stellar atmospheres and in some very dense star-forming gas.

The apparent difficulty of forming interstellar molecules is enhanced by the fact that they can be rapidly destroyed. Ultraviolet photodissociation occurs in a few hundred years for most molecules in interstellar regions unshielded by grains. We can see this using a typical photodissociation cross-section curve (see Figure 3.3) for H_2O. The figure shows cross-sections of approximately 10^{-21} m^2 occurring over a bandwidth of approximately 10 nm. Hence, a flux of ultraviolet photons of about 10^{10} m^{-2} s^{-1} nm^{-1} (which is the mean unshielded interstellar UV flux at these wavelengths) gives a lifetime of about 10^{10} s ($\approx$ 300 yr). In denser regions, molecules are shielded from UV radiation, but may be subject to various chemical reactions which may also effectively destroy them on a short timescale, and to ionization by cosmic rays.

At first sight, it is difficult to see how interstellar chemistry in cold neutral interstellar regions can proceed efficiently. In fact, to make this happen, the chemistry must be *driven* in some way. The two important drivers are starlight and cosmic rays: these drivers inject energy and create ionization. Other factors may also be important. Firstly, more frequent collisions would help: larger cross-sections may arise where electrostatic or induced forces play a part (see Section 3.4.2). Secondly, if, instead of a radiative recombination of atoms A and B as discussed earlier, we have a chemical reaction between *molecules*,

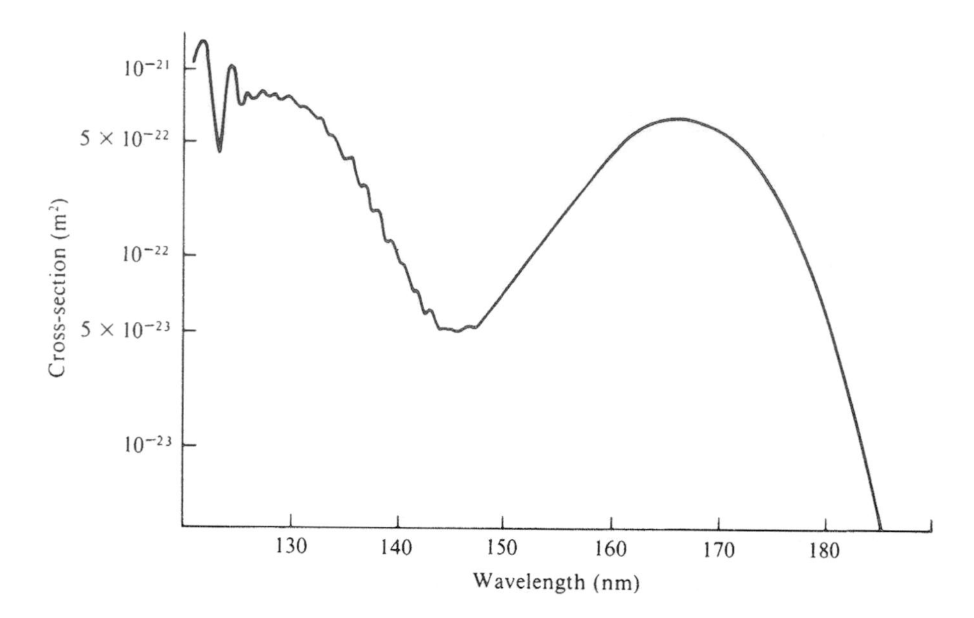

FIGURE 3.3 An example of the absorption of UV by a simple molecule: the figure shows the cross-section in m^2 per molecule of H_2O as a function of wavelength. The absorption of the radiation in most cases leads to dissociation to ground state products: $H_2O + h\nu \rightarrow H(^2S_{1/2}) + OH(^2\pi)$. The H atoms can be detected, and the quantum yield in this process is near unity.

$$A + B \rightarrow C + D, \tag{3.17}$$

then energy stabilization is not a problem, for the energy required to stabilize the products can appear as kinetic energy. Thirdly, if the time in contact, τ, for AB can be prolonged, then reaction may be more likely: this prolongation occurs naturally in the case of molecules containing at least a few atoms, because the energy in the system can be shared among a number of bonds so that the time in contact is increased and the possibility of radiating is enhanced. An alternative can be the effect of catalysis, which may occur on the surfaces of dust grains (we'll discuss surface chemistry in Chapter 4) or in interstellar ices (see Section 4.6.4). Finally, gas dynamics can create high temperature gas in shocks so that reactions that are suppressed at low temperature become viable (see Chapter 6).

3.4.2 Ion–Molecule and Neutral Exchange Reactions

Reactions between ions and molecules can be particularly efficient. For example, the reaction

$$O^+ + H_2 \rightarrow OH^+ + H \tag{3.18}$$

occurs with a rate coefficient of about 10^{-15} m^3 s^{-1} per molecule, implying a mean cross-section of approximately 10^{-18} m^2. Many such reactions are known to proceed rapidly, with about the same value for the rate coefficient. Why should this be? The physical reasons are rather interesting: the ion, in this case O^+, induces an electric dipole in the molecule, in this case H_2, and there is a resultant interaction energy of $\alpha e^2/8\pi\varepsilon_0 r^4$, where e is the charge on the ion, α is the polarizability of the molecule, and r their separation. According to the classical theory of orbits, the particles moving under this potential energy proportional to r^{-4} will spiral in towards each other if the impact parameter b is less than a critical value, b_0, dependent on the velocity, where

$$b_0 = \left(\frac{\alpha e^2}{\pi \varepsilon_0 \mu v^2} \right)^{1/4} \tag{3.19}$$

where v is the initial velocity and μ is the reduced mass (Figure 3.4). If $b < b_0$, then the particles will collide with considerable energy ($\approx$ 1 eV), enough to overcome almost any likely activation-energy barrier. The reactants then may rearrange themselves into the lowest-energy configuration, the excess energy appearing as kinetic. The cross-section for interaction is thus πb_0^2 and the rate coefficient is $\langle v\pi b_0^2 \rangle$, which we see from equation (3.19) is independent of velocity. Therefore, this simple theory predicts that all such ion–molecule reactions rates should vary only slightly with polarizability and with reduced mass, and be independent of temperature. Since most polarizabilities of molecules are of the same order of magnitude, the same order of magnitude for rate coefficients of exothermic ion–molecule reactions is to be expected. This is found to be the case.

Neutral exchange reactions between atoms and molecules or radicals may also occur in the interstellar medium if the reactions are exothermic. For example, the reaction

$$CH + O \rightarrow CO + H \tag{3.20}$$

is believed to occur with a rate coefficient of about 10^{-17} m^3 s^{-1} per molecule at room temperature, equivalent to a cross-section of about 10^{-19} m^2. In reactions like these, the energy of stabilization is carried away by the excess atom; in the case of reaction (3.20), the excess atom is H. The rearrangement favours the most strongly bound pair of atoms. In reaction (3.20), CH is bound by about 3.5 eV, whereas CO is bound by 11.1 eV. In these neutral reactions, however, the forces between the reactants are weak; at long range, they are merely van der Waals interactions. Therefore, we expect a smaller cross-section for neutral exchange reactions than for ion–molecule reactions. Many of the rate coefficients are of the same order as the one noted earlier, because van der Waals coefficients do not vary greatly from one molecule to another.

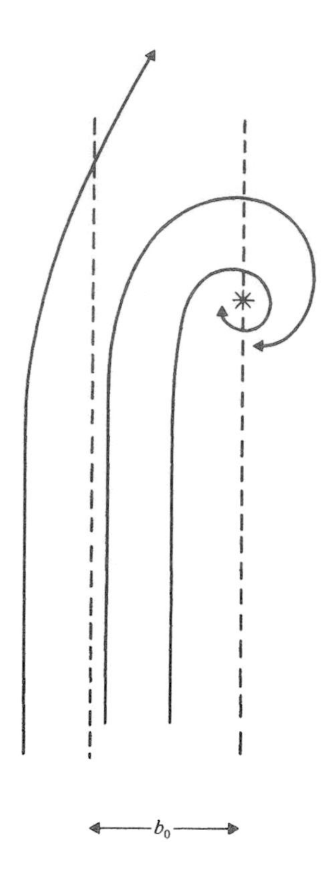

FIGURE 3.4 Interaction in ion–molecule reactions. If the impact parameter is less than b_0, interaction occurs.

Neutral reactions may also be impeded by activation energies. The reason for this can be understood on the basis of a simple picture. Figure 3.5 represents a collinear exchange reaction: A + BC → AB + C; this reaction forms a new molecule AB in place of the original molecule, BC. When the reactants are far apart, BC behaves like an independent molecule, oscillating with internal vibration. Similarly when the distance BC is large, the molecule AB oscillates in its potential well. The contour map suggests that the two 'valleys' meet at a 'pass' which is lower than the 'valley' walls, but still may represent a barrier to motion from one 'valley' to the next. While a reaction may have a large rate coefficient measured at room temperature, the value at a cloud temperature of around 20 K may be very different. A small activation energy of, say, 100 K equivalent energy gives a rate coefficient of a form proportional to $e^{-100/T}$. This would effectively suppress the reaction at temperatures much below 100 K, while allowing reactions above that temperature to occur. As we'll see in Chapter 6, shocks can heat interstellar gas and open up many chemical channels that are closed at the low temperatures of diffuse and dark clouds in the interstellar medium.

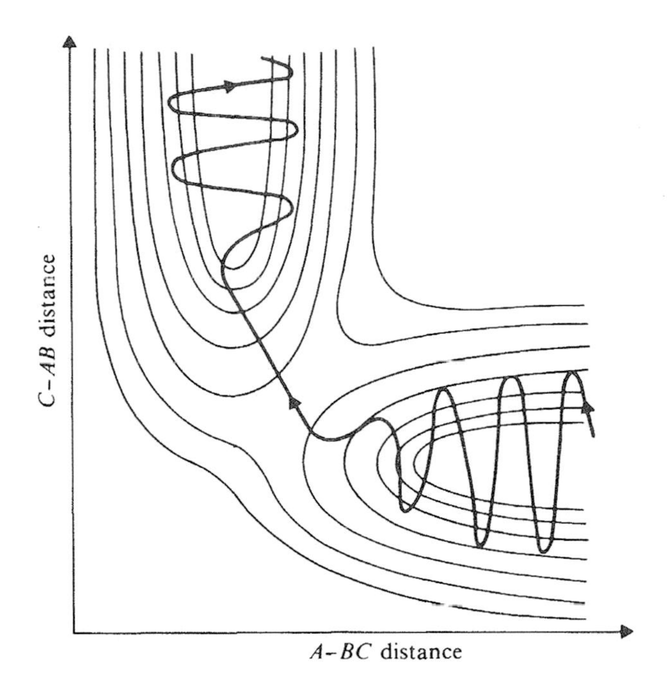

FIGURE 3.5 The diagram represents the interaction of A, B, and C as a contour map. Oscillation in the 'valleys' represents the molecules AB and BC. With sufficient kinetic energy, a particle may travel from one 'valley' to the other, or with lower energy penetrate the potential barrier by quantum mechanical tunnelling.

3.4.3 Gas Phase Reaction Networks

Ion–molecule and neutral exchange reactions are important contributors to the abundances of interstellar molecules, and networks containing thousands of such gas phase reactions (and other types) have been explored; they have had great success in describing the variety of observed molecules in different regions and their abundances, though some problems remain. These reaction schemes depend on the presence of an initial molecule, which is either H_2 or a molecule formed from H_2. The formation of this seminal species, that is, the molecule on which all of interstellar chemistry rests, will be discussed in the next chapter.

In studying reaction networks, we are hoping to describe the chemistry of interstellar clouds in sufficient completeness that we can calculate abundances of the molecule formed in the chemistry. Suppose we wish to calculate the number density $n(\mathrm{M})$ of a molecule, M, formed in the reaction

$$\mathrm{A} + \mathrm{B} \rightarrow \mathrm{M} + \mathrm{N}, \tag{3.21}$$

which has a rate coefficient k_f m^3 s^{-1} per molecule; the molecule M may be destroyed by photodissociation at a rate $\beta(\mathrm{M})$ per second and by reactions such as

$$M + X \rightarrow Y + Z, \tag{3.22}$$

which has a rate coefficient k_d m^3 s^{-1} per molecule. Then, we can write down a differential equation for n(M), showing how it changes in time, t,

$$\frac{d}{dt}n(M) = k_f n(A)n(B) - n(M)[\beta(M) + k_d n(X)] \tag{3.23}$$

where the first term on the right-hand side of equation (3.23) is the rate of formation of molecule M by reaction (3.21) (and all such reactions must be included), and the second term is the rate of loss of molecule M by photodissociation and photoionization and by reaction (3.22) with species X (again, all such reactions must be included).

An equation similar to equation (3.23) must be written down for each species included in the chemistry, so rate coefficients and photodissociation rates must be known (or estimated) for a large number of reactions. The physical parameters (number density, temperature, elemental abundances, radiation intensity, etc.) need to be specified or an evolving scenario must be defined. Then the equations can be integrated numerically. The process is simple in conception but may be complicated and laborious in practice. Extensive databases for astrochemistry are maintained and updated (see end of chapter), and codes for setting up and integrating the rate equations are available.

Sometimes, if the chemistry proceeds rapidly compared to any other changes, it may be sufficient to calculate the steady state abundances, that is, with $dn(M)/dt = 0$. To find the steady state n(M) from equation (3.23), we see that we need to know the number densities of all other species, A, B, X, and so on, so we have to write down a set of algebraic equations of the form of equation (3.23) for each atom, molecule, and ion in chemistry, but with each derivative on the left-hand side set equal to zero. Such sets of equations are solved numerically.

We can occasionally obtain useful information about steady state chemistry without solving equation but by considering a simple enough set of chemical equations. For example, in the centre of a dense cloud of molecular hydrogen, say $n(H_2) \gtrsim 10^{10}$ m^{-3}, the level of ionization is probably dominated by the cosmic ray ionization of H_2 at a rate ζ per second, the formation of H_3^+ in reaction (3.11) with rate coefficient k', and the dissociative recombination of H_3^+ with electrons in reaction (3.12) with a rate coefficient k m^3 s^{-1} per molecule. From the rate equations, we find

$$\zeta n(H_2) = k' n(H_2) n(H_2^+) = k n(H_3^+) n(e) \tag{3.24}$$

This equation allows us to predict that $n(H_3^+) > n(H_2^+)$ so that $n(H_3^+) \approx n(e)$, and we find

$$n(e) = \left(\frac{\zeta}{k} n(H_2)\right)^{1/2} \tag{3.25}$$

Typically, $\zeta = 10^{-17}\ s^{-1}$ and $k = 10^{-12}\ m^3\ s^{-1}$ per molecule, so that the fractional ionization is

$$n(e)/n(H_2) = 3 \times 10^{-3}/[n(H_2)]^{1/2}, \qquad (3.26)$$

which shows that the level of ionization to be expected in dense clouds is really very small. In clouds in which $n(H_2) = 10^{10}\ m^{-3}$, the ionization fraction predicted by equation (3.26) is only 3×10^{-8}. Although this treatment is greatly simplified, and a proper discussion should include other processes affecting the ionization, this theoretical result does seem to be approximately correct. In addition, observations of various molecular ions in dense clouds are consistent with a low level of ionization.

3.4.4 Initial Steps in Interstellar Cloud Gas Phase Chemistry

As we'll see in detail in Chapter 5, the radiation field from hot stars creates ionized regions (HII regions) around those stars. These HII regions trap all the stellar photons with energies greater than the ionization potential of hydrogen atoms, that is, 13.6 eV. This means that the spectrum of mean starlight outside those regions is truncated in energy, and possesses only photons with energies less than this value. However, the ionization potential of carbon atoms is 11.3 eV and of sulfur atoms is 10.4 eV, so in interstellar regions that are penetrated by starlight, these two elements (and some others) are photoionized, and are present in diffuse clouds as C^+ and S^+ ions. Molecular hydrogen is the most probable collision partner and at the low temperatures in diffuse clouds, C^+ ions and H_2 molecules radiatively associate to form CH_2^+, and successive reactions of this product with H_2 molecules leads to CH_3^+ and CH_5^+. Dissociative recombination of these ions with electrons forms a variety of simple hydrocarbons: CH, CH_2, CH_3, and CH_4. These species then undergo further reactions. For example, an exchange reaction of CH with an oxygen atom

$$CH + O \rightarrow CO + H \qquad (3.27)$$

forms the important molecule carbon monoxide, while greater molecular complexity can be achieved in reactions such as

$$CH_3 + O \rightarrow H_2CO + H \qquad (3.28)$$

leading to formaldehyde (a detected species in interstellar clouds). Evidently, a rich carbon chemistry may arise even in diffuse clouds. However, sulfur does not behave similarly to carbon, because the association of S^+ with H_2 is slow, so that sulfur chemistry in both diffuse and dark clouds grows not through sulfur hydride chemistry but mainly through exchange reactions with other species such as CH and OH, forming CS and SO.

In denser, dark clouds where starlight is excluded, carbon atoms are mainly neutral. Carbon chemistry proceeds through reaction with H_3^+ (formed in reaction (3.11)) to form CH^+ and subsequent hydrogen abstraction reactions with H_2 lead to CH_2^+ and CH_3^+, and a slower radiative association with H_2 provides CH_5^+. As in diffuse clouds, recombination reactions of these ions with electrons provide CH, CH_2, CH_3, and CH_4, and these molecules generate further chemistry.

Nitrogen and oxygen atoms have ionization potentials of 14.5 and 13.6 eV, respectively, so that they cannot be photoionized by starlight in the interstellar radiation field, and these species are mainly neutral in both diffuse and dark clouds. Oxygen atoms do not react with hydrogen molecules at the low temperatures of dark interstellar clouds, as there is too great a barrier of the type illustrated in Figure 3.5. However, oxygen can be chemically converted to other molecules even at low temperatures by reactions with the H_3^+ ion, produced through reaction (3.11). The H_3^+ ion can readily donate a proton to many species, including O atoms,

$$O + H_3^+ \rightarrow OH^+ + H_2 \tag{3.29}$$

and the OH^+ ion can react in a sequence of hydrogen abstraction reactions with H_2 molecules,

$$OH^+ \overset{H_2}{\rightarrow} OH_2^+ \overset{H_2}{\rightarrow} OH_3^+ \tag{3.30}$$

though the sequence stops there as O^+ has a valency of 3 and can bind no more H atoms. Dissociative recombination reactions with electrons produce OH and H_2O:

$$OH_3^+ + e \rightarrow \begin{cases} OH + H_2 \\ H_2O + H. \end{cases} \tag{3.31}$$

Both OH and H_2O are detected in dark clouds, and can undergo further ion–molecule reactions to produce other oxygen-containing molecules, including HCO^+ and CO:

$$C^+ + OH \rightarrow CO^+ + H \tag{3.32}$$

$$CO^+ + H_2 \rightarrow HCO^+ + H \tag{3.33}$$

$$C^+ + H_2O \rightarrow HCO^+ + H \tag{3.34}$$

$$HCO^+ + e \rightarrow H + CO. \tag{3.35}$$

However, the equivalent reactions of N with H_3^+ to form NH^+ or NH_2^+ do not proceed. An alternative entry to nitrogen chemistry exists: cosmic rays colliding with N atoms create N^+ ions, which react with H_2 molecules to form NH^+. Hydrogen abstraction reactions with H_2 molecules lead to the formation of NH_4^+, and dissociative recombinations of these ions with electrons give the species NH, NH_2, and NH_3. These species are then available for reactions with other atoms and ions, for example, to form products containing the CN radical.

Although the problem of generating chemistry in interstellar clouds seems at first difficult, once entry routes can be identified it is evident that enormous chemical opportunities become available. The schemes adopted are all broadly similar. In dark clouds, cosmic rays drive the formation of H_3^+ which donates protons to many species. Since the abundance of molecular hydrogen is overwhelming in interstellar clouds, if reactions of these ions can occur with H_2, they will do so; successive hydrogen abstraction reactions may occur (as in reactions (3.30)) followed by neutralization in reactions with electrons; this scheme creates hydrides and hydride ions which provide the feedstock for more complex reactions. The networks are essentially unrestricted, so that molecules containing more than just a few atoms can be readily produced in gas phase networks similar to those whose first steps are indicated here.

The molecular species that have been identified (by 2019) in many types of interstellar region (not only diffuse and dark clouds) are shown in Table 3.2. In fact, the actual number of interstellar molecular species must be much larger than the number included in this list of detected species (about 200 species), because the chemical networks to produce this variety of molecules include many species that have not yet been detected. The lack of detection of these species may be a result of low abundance or of the species not having a suitable transition for detection by current technology. Table 3.2 contains many molecules which have been detected in which the main isotopes have been replaced by minor isotopes. For example, H may be replaced by D, ^{12}C by ^{13}C, and ^{16}O by ^{17}O. These versions of the molecules are called isotopologues. Many of the species in Table 3.2 are also detected as isotopologues; if these are counted separately, then the total number of detected species is several times larger.

It is remarkable that very many of these detected interstellar species can be formed in simple but extensive gas phase schemes of the type outlined in this section. However, a significant number of exceptions exist, including the seminal molecule, H_2, and many of the largest molecules in the list. For these exceptions, it may be that appropriate gas phase chemical networks do not exist, or that those that do exist are incapable of providing particular molecules in the detected abundances. We shall describe an alternative to gas phase chemistry in the next chapter.

It is clear from this brief description that interstellar chemistry is dependent on the presence of molecular hydrogen. There are a number of ways in which interstellar molecular hydrogen can be formed in gas phase reactions but these ways are incapable of providing a fast enough formation rate, at least in the Milky Way galaxy. The direct association of two H atoms in the gas phase is very strongly forbidden by quantum mechanics and so does not contribute, and formation in exchange reactions such as $XH + H \rightarrow X + H_2$ is a circular argument since XH almost certainly requires H_2 for its formation. Two other possibilities remain. Hydrogen atoms may radiatively associate with electrons to form hydrogen negative ions, H^-, which then react with H atoms to form H_2 molecules, releasing the electrons back to the gas phase. The second possibility is a similar

TABLE 3.2

Detected interstellar molecular species (by 2019).

CH (methylidyne) **CN** (cyanide) **CH^+** (methylidyne cation) **OH** (hydroxyl) **CO** (carbon monoxide) **H_2** (molecular hydrogen) **SiO** (silicon monoxide) **CS** (carbon monosulfide) **SO** (sulfur monoxide) **SiS** (silicon monosulfide) **NS** (nitrogen sulfide) **C_2** (diatomic carbon) **NO** (nitric oxide) **HCl** (hydrogen chloride) **NaCl** (sodium chloride) **AlCl** (aluminium monchloride) **KCl** (potassium chloride) **AlF** (aluminium monofluoride) **PN** (phosphorus mononitride **SiC** (silicon carbide) **CP** (carbon monophosphide) **NH** (nitrogen monohydride) **SiN** (silicon mononitride) **SO^+** (sulfur monoxide cation) **CO^+** (carbon monoxide cation) **HF** (hydrogen fluoride) **N_2** (molecular nitrogen) **CF^+** (fluoromethylidynium cation) **PO** (phosphorus monoxide) **O_2** (molecular oxygen) **AlO** (aluminium monoxide) **CN^-** (cyanogen anion) **OH^+** (hydroxyl cation) **SH^+** (sulfur monohydride cation) **HCl^+** (hydrogen chloride cation) **SH** (sulfur monohydride cation) **TiO** (titanium oxide) **ArH^+** (argonium cation) **NS^+** (nitrogen sulfide cation) **HeH^+** (helium hydride cation)

H_2O (water) **HCO^+** (formylium cation) **HCN** (hydrogen cyanide) **OCS** (carbonyl sulfide) **HNC** (hydrogen isocyanide) **H_2S** (hydrogen sulfide) **N_2H^+** (protonated nitrogen) **C_2H** (ethynyl) **SO_2** (sulfur dioxide) **HCO** (formyl) **HNO** (nitroxyl) **HCS^+** (thioformyl cation) **HOC^+** (hydroxymethyliumidene cation) **SiC_2** (cylacyclopropynylidene) **C_2S** (dicarbon sulfide) **C_3** (tricarbon) **CO_2** (carbon dioxide) **CH_2** (methylene) **C_2O** (dicarbon monoxide) **MgNC** (magnesium isocyanide) **NH_2** (amidogen) **NaCN** (sodium cyanide) **N_2O** (nitrous oxide) **MgCN** (magnesium cyanide) **H_3^+** (protonated molecular hydrogen) **SiCN** (silicon monocyanide) **AlNC** (aluminium isocyanide) **SiNC** (silicon monoisocyanide) **HCP** (phosphaethyne) **CCP** (dicarbon phosphide) **AlOH** (aluminium hydroxide) **H_2O^+** (water cation) **H_2Cl^+** (chloronium cation) **KCN** (potassium cyanide) **FeCN** (iron cyanide) **HO_2** (hydroperoxyl) **TiO_2** (titanium dioxide) **CCN** (cyanomethylidine) **Si_2C** (disilicon carbide) **S_2H** (hydrogen disulfide) **HCS** (thioformyl) **HSC** (isothioformyl) **NCO** (isocyanate)

NH_3 (ammonia) **H_2CO** (formaldehyde) **HNCO** (isocyanic acid) **H_2CS** (thioformaldehyde) **C_2H_2** (acetylene) **C_3N** (cyanoethynyl) **HNCS** (isothiocyanic acid) **$HOCO^+$** (protonated carbon dioxide) **C_3O** (tricarbon monoxide) **l-C_3H** (propynylidyne) **$HCNH^+$** (protonated hydrogen cyanide) **H_3O^+** (protonated water) **C_3S** (tricarbon monosulfide) **c-C_3H** (cyclopropenylidene) **HC_2N** (cyanocarbene) **H_2CN** (methylene amidogen) **SiC_3** (silicon tricarbide) **CH_3** (methyl) **C_3N^-** (cyanoethynyl anion) **PH_3** (phosphine) **HCNO** (fulminic acid) **HOCN** (cyanic acid) **HSCN** (thiocyanic acid) **HOOH** (hydrogen peroxide) **l-C_3H^+** (propynylidyne cation) **HMgNC** (hydromagnesium isocyanide) **HCCO** (ketenyl) **CNCN** (isocyanogen) **HONO** (nitrous acid)

HC_3N (cyanoacetylene) **HCOOH** (formic acid) **CH_2NH** (methanimine) **NH_2CN** (cyanamide) **H_2CCO** (ketene) **C_4H** (butadiynyl) **SiH_4** (silane) **c-C_3H_2** (cyclopropenylidene) **CH_2CN** (cyanomethyl) **C_5** (pentacarbon) **SiC_4** (silicon tetracarbide) **H_2CCC** (propadienylidene) **CH_4** (methane) **HCCNC** (isocyanoacetylene) **HNCCC** (?) **H_2COH^+** (protonated formaldehyde) **C_4H^-** (butadiynyl anion) **CNCHO** (cyanoformaldehyde) **HNCNH** (carbodiimide) **CH_3O** (methoxy) **NH_3D^+** (deuterated ammonium cation) **H_2NCO^+** (protonated isocyanic acid) **$NCCNH^+$** (protonated cyanogen) **CH_3Cl** (chloromethane)

CH_3OH (methanol) **CH_3CN** (methyl cyanide) **NH_2CHO** (formamide) **CH_3SH** (methyl mercaptan) **C_2H_4** (ethylene) **C_5H** (pentynylidyne) **CH_3NC** (methyl isocyanide) **HC_2CHO** (propynal) **C_5S** (pentacarbon monosulfide) **HC_3NH^+** (protonated cyanoacetylene) **C_5N** (cyanobutadiynyl) **HC_4H** (diacetylene) **HC_4N** (?) **c-H_2C_3O** (cyclopropenone) **CH_2CNH** (ketenimine) **C_5N^-** (cyanobutadiynyl anion) **HNCHCN** (E-cyanomethanimine) **SiH_3CN** (silyl cyanide)

(Continued)

TABLE 3.2 (Cont.)

CH_3CHO (acetaldehyde) **CH_3CCH** (methyl acetylene) **CH_3NH_2** (methylamine) **CH_2CHCN** (vinyl cyanide) **HC_5N** (cyanodiacetylene) **C_6H** (hexatriynyl) **c-C_2H_4O** (ethylene oxide) **CH_2CHOH** (vinyl alcohol) **C_6H^-** (hexatriynyl anion) **CH_3NCO** (methyl isocyanate) **HC_5O** (butadiynylformyl)

$HCOOCH_3$ (methyl formate) **CH_3C_3N** (methylcyanoacetylene) **C_7H** (heptatriynylidyne) **CH_3COOH** (acetic acid) **H_2C_6** (hexpentaenylidene) **CH_2OHCHO** (glycolaldehyde) **HC_6H** (triacetylene) **CH_2CHCHO** (propenal) **CH_2CCHCN** (cyanoallene) **NH_2CH_2CN** (aminoacetonitrile) **CH_3CHNH** (ethanimine) **CH_3SiH_3** (methyl silane)

CH_3OCH_3 (dimethyl ether) **CH_3CH_2OH** (ethanol) **CH_3CH_2CN** (ethyl cyanide) **HC_7N** (cyanotriacetylene) **CH_3C_4H** (methyl diacetylene) **C_8H** (octatriynyl) **CH_3CONH_2** (acetamide) **C_8H^-** (octatriynyl anion) **CH_2CHCH_3** (propylene) **CH_3CH_2SH** (ethyl mercaptan) **HC_7O** (hexadiynylformyl)

$(CH_3)_2CO$ (acetone) **$HO(CH_2)_2OH$** (ethylene glycol) **CH_3CH_2CHO** (propanal) **CH_3C_5N** (methycyanodiacetylene) **CH_3CHCH_2O** (propylene oxide) **CH_3OCH_2OH** (methoxymethanol) **HC_9N** (cyanotetraacetylene) **CH_3C_6H** (methyltriacetylene) **CH_3CH_2OCHO** (ethyl formate) **CH_3COOCH_3** (methyl acetate)

C_6H_6 (benzene) **C_3H_7CN** (*n*- and *i*-propyl cyanide)
c-C_6H_5CN (benzonitrile)

Species are grouped by the number of atoms they contain. All of them are potential probes of the wide range of physical conditions in the interstellar medium. Many of these molecular species may be found not only in interstellar clouds but also in circumstellar regions. Most of them are detected by millimetre-wave and submillimetre-wave observations. Two other molecular species, the fullerenes (cage molecules) C_{60}^+ and C_{70}^+, are known to exist in the interstellar medium. There is also evidence for the existence of polycyclic aromatic hydrocarbon (PAH) molecules in the interstellar medium, but no precise identifications have been made. Almost all the data in this table have been extracted from B A McGuire, Astrophysical Journal Supplement Series 239 17 (2018)

sequence of reactions involving the radiative association of H atoms with H^+ ions to form H_2^+ ions, which then react with H atoms to form H_2 molecules, releasing the H^+ ions to the gas phase. These schemes are valid and operated to form the first molecules in the early Universe. However, under the physical conditions in the interstellar medium of Milky Way, the H_2 formation rate is inadequate to compete with the photodissociation by starlight described in Section 3.3.2. Another method of forming interstellar molecular hydrogen in the interstellar medium of the Milky Way galaxy is required, and is described in Section 4.6.2.

Chemical networks containing hundreds of species interacting in thousands of reactions are routinely used in astrochemical studies, creating a huge demand for accurate reaction rate data. Websites such as Kinetic Database for Astrochemistry (KIDA) and UMIST Database for Astrochemistry (UDfA) maintain extensive lists of assessed and updated data.

PROBLEMS

1. We shall define a cooling time t_c by $-T/(dT/dt)$. Consider a cloud in which $n(H) = 10^8\ m^{-3}$, $n(e) = 10^5\ m^{-3}$, and $n(C^+) = 4 \times 10^4\ m^{-3}$, at

a temperature of 200 K. Suppose that cooling occurs solely by the excitation of transitions in C^+ by collisions with electrons. Calculate the cooling time, t_c, in years.

2. Find the abundance of H_2 necessary in a cloud of density $n = 10^8\ m^{-3}$ at a temperature of 100 K to give a cooling rate due to H_2 equal to that due to C^+ and e, if $n(e) = n(C^+) = 10^4\ m^{-3}$.
3. Show that, in equilibrium, the heating rate due to photoionization of atom X is $\alpha n(e)n(X^+)E$, where α is the recombination rate coefficient for $X^+ + e \rightarrow X + h\nu$, and E is the excess energy released in the photoionization. What is the heating due to C atoms in the cloud of problem 1? (Use $\alpha = 10^{-17}\ m^3\ s^{-1}$.)

   Show that at $T = 15$ K (approximately) the heating and cooling due to C^+ balance *in any cloud.*
4. Find the fraction, f, of hydrogen in molecular form ($f = 2n(H_2)/n$) which will maintain a cloud temperature of 100 K by H_2 cooling in a cloud heated by photoelectric emission from grains at a rate $10^{-28}\ J\ m^{-3}\ s^{-1}$.
5. The molecule AB is made in the following ways: $A + B \rightarrow AB + h\nu$ and $A + BC \rightarrow AB + C$ and the rate coefficients are k_1 and k_2, respectively. AB is lost in the reaction $AB + D \rightarrow A + BD$ (rate coefficient k_3) and in photodissociation $AB \rightarrow A + B$ (rate β). Write down an expression for the equilibrium abundance of AB in terms of other abundances.
6. Suppose that the chemistries of H_2, H_2^+, and H_2^+ are described by

$$H_2 + \text{cosmic ray} \rightarrow H_2^+ + e, \qquad \xi = 10^{-17} s^{-1}$$

$$H_2 + H_2^+ \rightarrow H_3^+ + H, \qquad k_1 = 10^{-15} m^3 s^{-1}$$

$$H_3^+ + CO \rightarrow HCO^+ + H_2, \qquad k_2 = 10^{-15} m^3 s^{-1}$$

   If $n(CO) = 10^{-4} n(H_2)$ and $n(H_2) = 10^{10}\ m^{-3}$ (constant), find the steady state abundances of H_2^+ and H_3^+.
7. For the chemistry of problem 6, write down the differential equation that determines the time dependence of H_2^+. Show that $n(H_2^+)$ approaches steady state on a timescale $1/[k_1 n(H_2)]$. If the typical lifetime of a cloud is 10^6 years, is steady state a good approximation for the abundance of H_2^+?

4 Interstellar Grains

4.1 EVIDENCE FOR INTERSTELLAR DUST GRAINS

4.1.1 DEPLETION OF ELEMENTS

The relative abundances of elements in atmospheres of hot stars may be measured and assumed to be a standard set of relative elemental abundances for the Galaxy. Relative elemental abundances measured in interstellar clouds can be very different from the standard set: relative abundances of some elements measured in diffuse clouds can be very greatly reduced – or *depleted* – in comparison to the standard set (see Table 4.1). Where have all these atoms gone? The elements which are capable of forming refractory (that is heat-stable and resistant) solids are among those which have high depletions. For example, silicon and oxygen form silicates with magnesium and iron. Iron may also form solid particles; silicon carbide is very stable. These data suggest that, in gas that is cooling as it moves away from a star, some elements that can form refractory solids are removed from the gas as solid particles, that is, dust grains.

4.1.2 EXTINCTION

In Chapter 2, we described observations which suggest that there is an agent in the Galaxy tending to obscure the stars and other sources of radiation: we called this property *extinction.*

We can define an extinction coefficient α such that the intensity I_0 of a star is reduced to

$$I = I_0 \exp\left(- \int_0^l \alpha \, dl \right) \tag{4.1}$$

after passage through a distance l of the interstellar medium. The coefficient α is not constant, but seems to be proportional to the density of gas in the interstellar medium, and the density itself may vary with position.

We can measure extinction most easily by comparing two similar stars (i.e. having similar masses, temperatures, and compositions, as revealed by their spectra), one of which has little material between it and us and the other being behind an interstellar gas cloud. Differences in the intensities received from these two stars may derive in part from differences in distance: this can be allowed for; but there will also be a difference because of extinction. Astronomers measure intensity changes on a logarithmic scale in units of magnitudes. These units are

TABLE 4.1

The measured relative abundances of some elements in interstellar gas and interstellar dust are given relative to one million H atoms.

Element	Carbon	Nitrogen	Oxygen	Magnesium	Silicon	Iron
Standard abundance	214	62	575	36.3	31.6	33.1
Abundance in IS gas	91	62	389	1.5	2.2	0.3
Percentage in IS gas	43%	100%	68%	4%	7%	0.9%
Abundance in IS dust	123	0	186	34.8	29.4	32.8
Percentage in IS dust	57%	0	32%	96%	93%	99.1%

The standard abundances are measured in the atmospheres of hot stars in which dust grains cannot survive, so that the atoms of which they are composed must be in the gas phase. The interstellar (IS) abundances are measured in diffuse clouds in which some of the elements are assumed to be included in dust grains and therefore no longer in the gas phase. The abundance of elements in dust grains is inferred from the difference between the two measured abundances. These data imply that in some cases, almost all of a particular element may be locked in dust grains (e.g. Mg, Si, and Fe). Significant amounts of carbon and oxygen remain in the gas and available for chemistry, while substantial amounts of these elements are also locked in dust. Nitrogen does not contribute to the dust component.

such that a change in intensity by a factor of 100 corresponds to a magnitude difference of 5; we can, therefore, relate two intensities I_1 and I_2 to a magnitude difference Δm by

$$\Delta m = 2.5\log_{10}(I_1/I_2). \tag{4.2}$$

Objects with smallest magnitudes are brightest. In these units, the typical extinction in the visible regions of the spectrum found in the thin gas filling most of the Galaxy is about 1 magnitude for a path length of 1 kpc. But in denser clouds, of course, the extinction is very much greater, and such clouds appear dark against the background of stars. Indeed, number counts per unit area on the sky of stars exceeding a certain brightness give us another general method of estimating interstellar extinction. Figure 4.1 shows the average interstellar extinction curve for the Milky Way galaxy. It has a characteristic shape, rising from the infrared through the visible region to a prominent 'bump' at 220 nm in the near ultraviolet. The curve then recovers and rises strongly into the far ultraviolet.

Extinction can, in principle, be measured at any frequency or wavelength, and observations have been made at wavelengths from the far UV ($\lambda \approx 100$ nm) through the visible ($\lambda = 0.4$–0.7 μm) to the near IR ($\lambda \approx 1$–20 μm). Extinction is found to vary with wavelength in the way indicated in Figure 4.1. Some differences are found in different regions of the Galaxy; however, the general shape of the extinction curve is usually as indicated. The most pronounced features are the

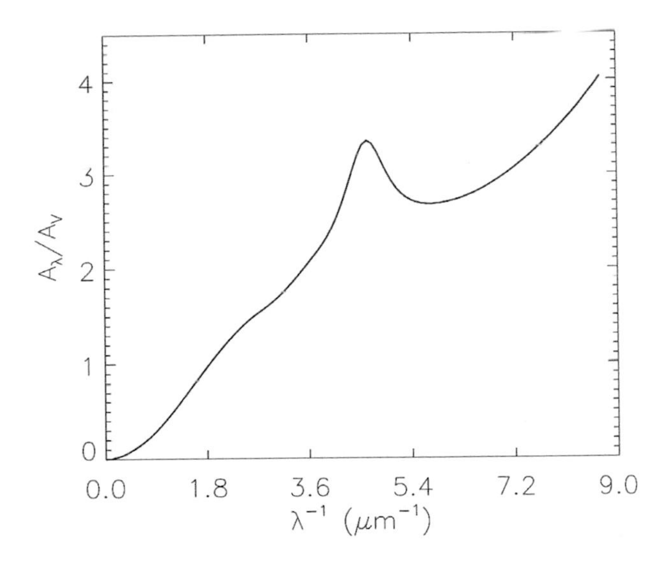

FIGURE 4.1 The average interstellar extinction curve for the Milky Way galaxy. The average extinction at wavelength λ, A_λ, is shown normalized to the extinction at wavelength V = 550 nm, A_V, and is plotted in this diagram as a function of $1/\lambda$ where λ is measured in μm. Therefore, the infrared is to the left and ultraviolet to the right, in this diagram. (Credit: C Cecchi-Pestellini.)

peak at $\lambda \approx 220$ nm or $1/\lambda \approx 4.6\ \mu m^{-1}$, and the strongly rising portion in the far UV. It is also particularly important to notice that in the visible region of the spectrum, the extinction is approximately proportional to $1/\lambda$; this near-linear portion of the curve first gave a clue as to the origin of extinction. As we discussed in Chapter 3, extinction is thought to be caused by small solid particles called dust grains which have diameters comparable to the wavelength of visible light. We shall show in Section 4.2.1 that such particles may be expected to produce the $1/\lambda$ behaviour, but that other components in the grains may be necessary to explain other features in the curve. Before we do that, we shall review some other evidence for the existence of grains in the interstellar medium.

4.1.3 Polarization

Although extinction points strongly to the existence of interstellar grains, one might feel that with sufficient ingenuity, an alternative source of extinction could be found. However, the discovery that starlight is – in general – partially linearly polarized made such a position more or less untenable. Observations of starlight show that visible light is often linearly polarized by a few per cent. The amount of polarization observed seems to be proportional to the amount of extinction.

To create in the laboratory a beam of partially polarized light travelling, say, in the z-direction, we use a polarizing agent such as a piece of polaroid plastic. This is a material which is stressed so as to produce an alignment of the molecules in the plastic material. The result is that the electric vector in one direction (say the x-direction) is preferentially absorbed with respect to the electric vector in the y-direction. This mechanism of polarization is therefore one of preferential extinction (but others also exist).

If this is the mechanism at work in the interstellar medium, then we require two things: first, interstellar dust grains of isotropic behaviour cannot be spherical, but must be elongated; and second, there must be some degree of alignment of these elongated grains. If these two criteria are satisfied, then radiation with electric vectors parallel to the longer axes of the grains will be more heavily extinguished than vectors parallel to shorter axes and polarization occurs. Aligned grains of anisotropic material (such as graphite) may also cause polarization. All dust grains will normally be rotating; if there is equipartition of energy between gas and grains, then

$$\frac{1}{2}I\omega^2 = \frac{3}{2}kT \tag{4.3}$$

for a grain of moment of inertia I rotating with angular frequency ω in a gas at temperature T. For grains of radius $a \approx 10^{-7}$ m and density 10^3 kg m^{-3}, we expect frequencies of rotation of the order of 10^5 Hz. What we require is a partial alignment of the axes of rotation; we require them to point preferentially in the same direction. We now describe one possible alignment mechanism.

Suppose that a magnetic field exists in the interstellar gas and that the grains are paramagnetic. Then the field induces a magnetic moment within the grain, which is continually changing its orientation within the grain, due to the grain's rotation. The continual change of the magnetic moment inside the grain requires the expenditure of energy which must come from the kinetic energy of rotation of the grain. The energy appears as heat, and the efficiency of the process depends on the imaginary part of the magnetic susceptibility. The drag on the motion is greatest when the grain is rotating about an axis perpendicular to the direction of the external field, and least when the axis is parallel to the external field. Thus, the former motion tends to be damped out and some measure of alignment is achieved. This effect, of course, is opposed by the effects of collisions which tend to randomize the spin axes of grain, so the degree of alignment is a subtle balance of the magnetic field and the magnitude of the imaginary part of the susceptibility against the effects of density and temperature. Collisions will succeed in randomizing the orientation unless the gas and grain temperatures are different.

4.1.4 Scattered Light

There is another way in which the grains betray their presence, and it is one which would be difficult to understand except as a property of dust grains.

The Galaxy is filled with a *diffuse light*, not directed from any particular source. The most natural origin for this is in the scattering of starlight by some agency. The diffuse light does not appear to have its origin in any other physical process. Scattering by atoms and molecules would be quite inadequate to account for it. The cross-section for Rayleigh scattering of light (i.e. without change in wavelength) by atoms is approximately 10^{-28} m^2, so the scattering effect of H atoms along the path is roughly measured by $10^{-28}n(\mathrm{H})$. For spherical dust grains of radius a and number density n_g, as we shall see, the equivalent measure for scattering by grains is approximately $\pi a^2 n_g$ in the visible, which is several orders of magnitude larger than for scattering by atoms. We infer that the diffuse light arises because of scattering of starlight by grains. It should, therefore, contain information about the grain properties: in particular, the scattered light contains information about the 'albedo' (or reflectivity) and also the 'phase factor', which describes whether grains scatter light preferentially forwards or backwards. We shall see in Section 4.2 that this information is complementary to that contained in the extinction observations.

Besides the general diffuse light, there are some places in which scattered light is directly observed as a reflection nebula. The cluster of stars known as the Pleiades in the constellation Taurus contains stars possessing 'halos'. These stars are situated sufficiently near to a gas and dust cloud that the scattered light is intense enough to be seen even by naked-eye observation as haziness around the stars in the cluster (see Figure 4.2).

4.1.5 Solid State Spectral Lines

There are several infrared absorption lines that are attributed to absorption of radiation from background stars by foreground solid particles (see Figure 4.3). There is a broad absorption feature with central wavelength at about 9.7 μm, which is about 1 μm wide at half intensity. This is attributed to the Si–O bond in amorphous (i.e. non-crystalline) silicates such as magnesium silicate, Mg_2SiO_4, or iron silicate, $FeSiO_4$, or some mixture of these and other silicates. The feature is caused by stimulating the stretching mode in this bond, and is broadened by interactions of the Si–O structure with its environment. The Si–O is not able to rotate, as a free SiO molecule would, so lines associated with SiO rotation (see Chapter 3) are absent. However, the SiO structure can also bend to and fro, relative to the rest of the molecule, and setting up this oscillation – which requires less energy than exciting the SiO stretching modes – causes absorption at wavelengths of about 18 μm. This absorption is also detected in the interstellar medium.

In dark regions, where the dust causes extinction of several magnitudes at visual wavelengths, a very strong absorption of the light of background stars is usually detected at a wavelength of about 3.0 μm. This is attributed to excitation of the O–H stretch mode in amorphous H_2O ice. In some sources, much of the oxygen is tied up in this form. In addition, particularly dark regions often show absorption from CO ice, at wavelengths near 4.7 μm. This

FIGURE 4.2 Scattered light near a star in the Pleiades cluster. Dust in a cloud near the star scatters starlight towards Earth. The scattered light shows that the cloud has filamentary structure – see Chapter 8. (Credit and copyright: Adam Block/Mount Lemmon SkyCenter/University of Arizona. Licensed under Creative Commons Attribution-Share Alike 4.0 International license.)

corresponds to the excitation of the CO vibration $\upsilon'' = 0 \rightarrow \upsilon'' = 1$. We can be sure that this is not arising in the gas phase CO because the associated rotational structure from free molecules of CO (see Chapter 3) is absent. This solid CO feature is rather sensitive in its central wavelength and width to the environment of the CO; pure solid CO has a narrower feature slightly shifted from that due to a solid solution of CO in H_2O. Therefore, infrared spectroscopy of solids is a useful tool, although not quite as specific in its assignments as atomic spectroscopy. The absorption features arise in particular bonds of a molecule, and may not define that molecule completely. However, details of the spectrum usually show the influence of the atomic-scale environment in which the bond is found.

If dust grains are hot enough, they will emit radiation not only in a continuum but also in spectral lines. Obviously, H_2O ice would evaporate before it could

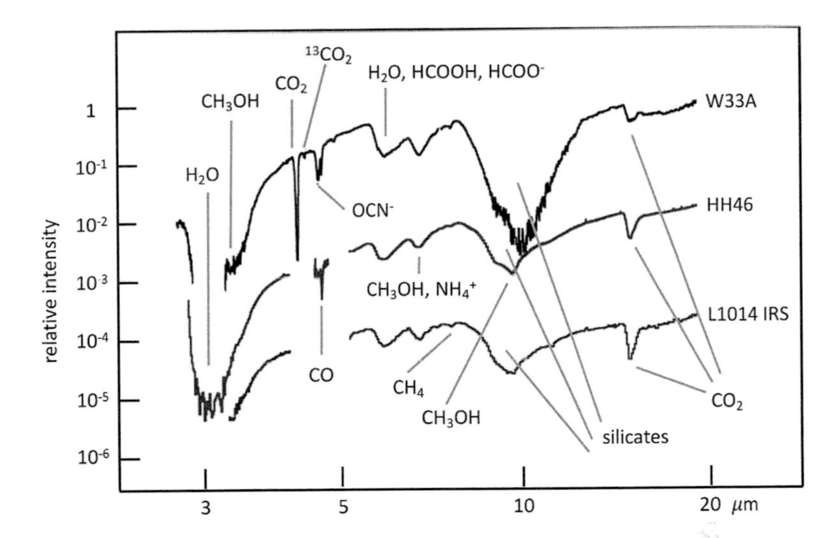

FIGURE 4.3 IR absorption spectrum of interstellar dust along lines of sight through dense interstellar clouds towards three stars of the Milky Way. The data are taken from K I Öberg et al., Astrophysical Journal 2011 740 109. Absorptions by a variety of molecular species in solid dust grains are evident in these spectra. (Credit for figure: C Cecchi-Pestellini.)

radiate at 3 μm (emission at this wavelength requires a temperature of about 1000 K), but silicates are much more robust at high temperatures, and the 9.7 μm feature can be detected in emission from dust that is heated to temperatures above several hundred K by radiation from a nearby star.

Other emission features between 3.3 and 11.3 μm are detected in some hot sources, and are attributed to hydrocarbon material that may be either in grains small enough to be made hot by a single UV photon, or to free-flying hydrocarbon molecules (PAHs). The 3.3 μm feature is associated with a C–H stretch mode, and the 11.3 μm feature with a C–H bend mode. Neither is specific to a particular species, and so we can use this information only to confirm the existence of hydrocarbon material in interstellar space.

4.1.6 Luminescence from Grains

A broad band emission centred on the red or near infrared and several hundred nm broad is observed from certain interstellar regions of high excitation. This emission is interpreted as bandgap luminescence emission from a material that has a bandgap energy $E_g \sim 2$ eV (see Figure 4.4). Possible materials with E_g in this range are H-rich amorphous hydrocarbons. Absorption in the UV and visible from the valence band leads to the population of states in the conduction band. Leakage from these conduction band states

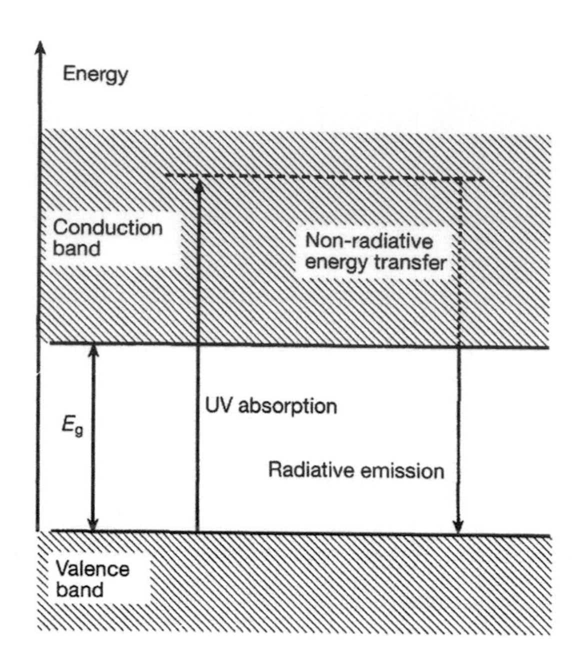

FIGURE 4.4 UV absorption by a solid may lead to emission of radiation with photons of energy around the bandgap energy, E_g.

into other modes occurs, until emission across the bandgap occurs, into the valence band.

4.1.7 Other Evidence for Dust Grains

All the evidence presented so far deals with dust remotely and therefore describes dust in an average way. However, it is now possible to identify single interstellar dust grains in primitive meteorites and in cometary dust that has been captured and brought to Earth. While most dust in the solar system has been radically affected by the processes that formed the Sun and planets, a small fraction of dust came through those events unchanged. It is possible to determine the origin of these 'pre-solar' grains from the isotopic abundances of the elements that they contain, as those isotopic abundances vary widely from one source (e.g. supernovae, novae, cool stellar winds, etc.) to another. Therefore, information about individual dust grains, rather than averages for dust grains on long paths through the interstellar medium can now be obtained. This information shows that the detailed picture of interstellar dust, in terms of sizes and of variety of materials, is much more complex than the rather simple picture emerging from the Sections 4.1.1–4.1.6. Nevertheless, the bulk properties of interstellar dust are best described by remote observation.

4.2 OPTICS OF DUST GRAINS

4.2.1 INTRODUCTION

The major problem concerned with the dust grains is to identify the materials of which they are made and the distributions of their shape and size. We may then go on to interpret the observations described earlier and to predict the effects the grains may have on the gas surrounding them. To help us in this study, we have the pieces of evidence described in Section 4.1. We must, in particular, be able to describe the optics of grains: how dust grains of a given size, shape, and material (and, therefore, refractive index) modify the electromagnetic radiation which falls on them. The problem is, in a sense, a quite straightforward task in classical physics: it involves the application of Maxwell's equations of electromagnetism to the interaction of a beam of light with a grain, with proper attention to the boundary conditions at the surface of the grain. In practice, although the purpose is simple to state, it is a complicated analytical and computational problem, especially in the case in which the size of the grain and the wavelength of incident light are comparable. The solution is named after its originator, Gustav Mie, and is now available for grains such as single spheres, concentric spheres, and clusters of spheres, and infinite cylinders.

Let us assume that there is a cloud of grains, each of the grains having radius a, with uniform number density n_g (m^{-3}). Then, in the case of single scattering, equation (4.1) is replaced by

$$I = I_0 \exp\left(-n_g \pi a^2 Q_{ext} l\right). \tag{4.4}$$

Q_{ext} is called the efficiency factor for extinction, and is made up of two parts

$$Q_{ext} = Q_{abs} + Q_{sca} \tag{4.5}$$

due to absorption and scattering. When calculating extinction, we must allow for all photons lost from the beam, so both absorption and scattering must be included. Since we are assuming only single scattering, the scattered photons are never put back into the beam. There are, of course, situations where multiple scattering is important, but we are not concerned with them here. How much scattering and how much absorption occurs depends, ultimately, on the refractive index, m, of the grain material. The refractive index is, in general, complex, and it is the imaginary part that gives rise to absorption, as we can see from the following argument. Suppose an electromagnetic wave with an electric field, E, of the form $E \propto e^{i(kx-\omega t)}$ passes along the x-axis through a medium in which m is complex. The wave has velocity c/m, or in our notation ω/k; thus, since m is complex, k must be complex, say, $k = k' + ik''$ where k' and k'' are real. In other words, the wave has a form $E \propto e^{i(k'x-\omega t)}e^{-k''x}$, which shows clearly that the amplitude of the wave decays exponentially with distance penetrated into the medium.

Results for the computed behaviour of Q_{ext} as a function of the dimensionless parameter $x = 2\pi a/\lambda$ for specific values of the refractive index show an initial near-linear rise, as x increases through values on the order of unity, similar to the rise in interstellar extinction curve (Figure 4.2) as a function of $1/\lambda$. This similarity was the original support for the hypothesis of interstellar grains. It is interesting to note that for large x, say $x \sim 20$, Q_{ext} approaches 2; that is, a grain that is large compared to the wavelength removes from the incident light exactly twice the amount of light it can intercept. This surprising result arises because we have assumed that *any* light scattered, no matter at how small an angle, is lost from the beam, and that the grain is very far away from us so that it does not cast a direct shadow.

Early studies of interstellar extinction were confined to the optical part of the spectrum and a restricted range of grain sizes was adequate to explain the optical extinction. The more complex structure of the interstellar extinction curve shown in Figure 4.1 requires a wide range of grain sizes and several components of different materials. The curve shown is an average for several lines of sight in the Milky Way. As we discuss in the next section, every line of sight to a star in the Milky Way has its own specific extinction curve (although generally similar), indicating that the size and size range and materials of interstellar dust grains may vary throughout the interstellar medium.

4.2.2 Application to the Interstellar Medium

The aim is to construct a curve such as that in Figure 4.1 by choosing 'correctly' a size distribution for the grains and the refractive index of the material of which they are composed, consistent with the information on elemental depletions. In fact, the problem is over-determined, and by suitably juggling with the parameters, one may hope to produce a 'correct' fit to any portion of the extinction curve. Unfortunately, this is not a guarantee of uniqueness.

This is why the other observational evidence (Section 4.1) is so important. The constraints placed by the total observational evidence are tight and we are forced to conclude that there must be several components in the grains, each responsible for different aspects of behaviour. Thus, for example, small carbonaceous grains (or large molecules) may be responsible for the extinction 'bump' near 220 nm in the UV. However, these grains cannot account satisfactorily either for the visible or the far-UV extinction. Very small grains have been suggested as the explanation for the far-UV extinction, and somewhat larger grains cause much of the visible extinction. A conventional assumption about grains is that they have radii $a = 3 \times 10^{-7}$ m, their number density n_g is about $10^{-12}n_H$, and their composition is probably some kind of silicate. These numbers can reproduce the visual extinction satisfactorily, and we shall sometimes use them for illustrative purposes. But a distribution of grain sizes and grain materials is certainly necessary; a common assumption is that the number in the size range $a \to a + da$ is proportional to $a^{-3.5}\ da$ so that there are many more small grains than large grains. Models of grains obtained in this way generally have size ranges from a few tenths of one micrometre for the large grains, down to a few nanometres for the smallest grains. Commonly adopted materials, consistent with elemental depletions, are silicates of various kinds and carbons. Some models adopt, in addition to solid dust particles, a population of PAH molecules. With assumptions like these, essentially perfect fits from the infrared to the ultraviolet to the observed interstellar extinction curves along any line of sight in the Galaxy may be obtained, and these fits are generally consistent with depletion data.

4.3 FORMATION AND DESTRUCTION OF GRAINS

4.3.1 Formation of Dust Grains

Where can the grains be formed? We might first think that they grow slowly in typical interstellar clouds, where we see them now. However, growth of grains in interstellar clouds takes a very long time, as we shall see by the following simple argument. Suppose at time $t = 0$, the grain radius is $r(0)$, and that the grain grows by the addition of species i (an atom or molecule) which has mass m_i and mean thermal velocity $\bar{v}_i$. Let s be the bulk density of the material. Then, at time t, we easily find that

$$r(t) = r(0) + \frac{\epsilon n_i m_i \bar{v}_i}{4s} t, \tag{4.6}$$

where ϵ is a sticking coefficient, between zero and unity. If small condensation centres exist, then the minimum time required to grow a solid of density $s = 10^3$ kg m^{-3} to radii $\approx 10^{-7}$ m is about $10^{20}/\epsilon n_i$ seconds, which for $n = 10^7$ m^{-3} and $n_i < 10^3$ m^{-3} exceeds $3 \times 10^9/\epsilon$ yr. Since $\epsilon \lesssim 1$, this is a long time compared with other relevant timescales, and may be comparable with the age of the Universe. Clearly, this is far too long and we must look instead for much denser regions where timescales for dust formation should be shorter. High densities are found in the outflowing gas (or stellar winds) from cool stars and there is strong observational evidence for dust formation in stellar winds. Dust is also observed to form in the ejecta from supernovae and novae, even though at first sight these are apparently rather hostile environments.

A striking example of a dust-forming stellar wind is the variable star R Corona Borealis whose perceived intensity in the visual part of the spectrum is typically fairly constant for years but from time to time abruptly declines by a large factor, only to recover over a period of a few months. This behaviour is now accepted as being caused by short periods of rapid dust formation in the extended envelope of the star. The formation of the dust extinguishes almost all of the light from the embedded star. However, as the envelope then drifts away from the star, its volume expands and the number density of dust grains per unit volume declines. The extinction caused by dust in the envelope is therefore reduced, and so the perceived stellar intensity recovers. Thus, in objects like R Corona Borealis, we are seeing dust formation and dissipation occurring, essentially in real time. Many low mass stars show a similar behaviour towards the ends of their lives.

Stellar winds associated with low mass cool stars are simply the stellar atmospheres that have been set in motion, possibly by stellar pulsations, and so the starting conditions that we need to understand for stellar wind chemistry and dust formation are simply the conditions in the stellar atmospheres. The conditions in the atmospheres of cool stars – number densities of about 10^{19} H nuclei m^{-3} and temperatures of about (1–2) $\times$ 10^3 K – are clearly very different from those that apply in diffuse and dark clouds in the interstellar medium and these atmospheres are entirely molecular. There are two types of situation to consider: as a result of differences in stellar evolution, cool stars are found to be either oxygen-rich or carbon-rich, and the chemistry in the stellar atmospheres differs in the two cases.

The equilibrium gas phase state in a stellar atmosphere for elemental abundances and specified temperature and pressure is readily obtained. This state corresponds to the minimum of the thermodynamic Gibbs function for all the species involved in the stellar atmosphere. A set of algebraic equations for the partial pressure of each species can be set up and solved iteratively for any given temperature and pressure.

In the O-rich case, where the abundance of oxygen in the stellar atmosphere exceeds that of carbon, these calculations show that almost all the carbon is locked in carbon monoxide and any other carbon-containing species have a very low abundance. The oxygen that is not in CO is mainly in water and silicon monoxide. Aluminium and calcium appear in a variety of oxides, while iron and magnesium are mainly atomic. Since CO is an unlikely component of dust, and taking account of elemental abundances, these studies lead to the conclusion that in O-rich conditions, dust may form from water, atomic iron, atomic magnesium, and silicon monoxide.

In the C-rich case, the results show that almost all the oxygen is locked in CO, and the chemistry of the excess carbon is dominated by atomic carbon above 1700 K, by C_2H at 1600 K, and by C_2H_2 at lower temperatures. The metals iron and magnesium are atomic, as is silicon at high temperatures. In winds arising from these atmospheres, the dust is likely to be amorphous carbon and some metallic dust may also appear.

The generally accepted view of dust formation is that the growth of dust grains takes place by condensation from the gas on to nuclei that have formed earlier in the outflow. These nuclei may not have the same chemical composition as the material that condenses on to them. In this scenario, a limiting step in dust formation is the creation of nuclei on which condensation of species may occur to form the dust from the main species. What are these nuclei? Fortunately, it is possible to calculate, for a given temperature and pressure, from the partial pressures obtained in the gas phase calculations and from their thermodynamic free energies whether nucleation clusters condensing in the gas are likely grow or decline. These clusters will be formed of the molecular species that appear earliest as the gas cools. As we have seen earlier, in O-rich winds, iron, magnesium, and silicon, and molecules such as Mg_2SiO_4

and $MgSiO_3$ should eventually form dust. However, species of lower abundance can condense at high temperatures during the cooling process. These are insufficiently abundant to provide the entire dust population but may provide the nucleation centres on which the more abundant species can condense at higher temperatures during the cooling process. These low abundance nucleation clusters include species such as Al_2O_3, TiO_2, $CaTiO_3$, and ZrO_2. Once the nucleation centres are available, then the abundant species such as SiO, Mg, and H_2O, for example, may combine on the surface to form solid Mg_2SiO_4. In C-rich winds, carbon is able to form solid soot from C_2H_2 even at the highest temperatures in the cooling process. Radiation pressure from the central cool star on the newly formed dust grains helps to drive the outflow in the stellar envelope. Outflow velocities are observed to be ~10 km s^{-1}. Velocities as large as this can be achieved by radiation pressure on dust.

Supernovae are also important sources of interstellar dust, in fact, they are even more important than dust formation in the winds of cool stars. Indeed, in the early Universe (up to a billion years after the Big Bang), supernovae were the only sources of dust because the low mass cool stars had not evolved by that time into a dust-forming stage. Supernovae are stellar explosions that may abruptly reach a maximum brightness comparable to that of an entire galaxy. The brightness then fades over a period of some weeks. Over the whole event, a supernova may emit as much energy as a star like the Sun emits in its whole life. Supernovae may occur in stellar binaries when matter is rapidly transferred from a stellar giant to a condensed star known as a white dwarf. The abrupt increase in mass of the white dwarf compressed under its intense gravity may trigger a temperature rise sufficient to initiate nuclear reactions that synthesize heavy elements and drive the stellar explosion. This type of supernova is called Type Ia. Another type of supernova (Type II) occurs in a single massive star when the normal sequence of nuclear reactions initially converts hydrogen to helium, then to carbon, nitrogen, oxygen, silicon, and ultimately to iron. These products may form a shell structure within the star. However, the sequence of nuclear reactions ceases when iron is formed, so that the energy source in the star abruptly terminates. The star implodes on to the core and rebounds, sending matter that was in the star, including radioactive isotopes (especially ^{56}Co) and neutrinos, into the surrounding space. Emission of γ-rays from ^{56}Co helps to power the brightness variation of the supernova but generates fast particles and a harsh UV field intrinsic to the gas; these processes tend to inhibit dust formation in the ejecta. Supernovae of Type II are generally fainter than those of Type Ia but detailed studies suggest that Type II supernovae contribute significantly more dust than those of Type Ia.

In the simplest view, the layers of H, He, C, N, O, Si, and Fe in Type II core collapse supernovae remain distinct. However, if some mixing occurs, then the opportunity for chemistry and dust formation in the ejecta increases. Current models show that dust formation should occur, and that a few tenths of a solar mass should be formed. Observations of SN1987a, a supernova in the neighbouring galaxy to the Milky Way, the Large Magellanic Cloud, show that molecular rovibrational emissions from CO and SiO appeared at about one hundred days after outburst, and infrared continuum emission from solid dust grains appeared after about one year. The amount of dust was initially very small, but after a few decades elapsed, the amount of dust formed rose to almost one solar mass. The ejecta from SN1987a are moving at high speed (about 10 000 km s^{-1}) but has not yet impacted on material surrounding the supernova. The reverse shock created by the impact will generate shock temperatures of a million K and create a so-called supernova remnant (see Figure 1.3). The size distribution and chemical nature of dust from SN1987a will be affected by the supernova remnant.

Cool stellar envelopes and supernovae are not the only sources of dust grains, but they are the most prolific. Other sources include planetary nebulae, novae, and some special hot stars. It seems that objects within the Universe have a tendency to produce dust, even in circumstances that are apparently hostile. As we'll see, dust grains play important roles in the evolution of the interstellar medium.

4.3.2 Dust Evolution and Destruction in the Interstellar Medium

Dust grains formed in the near-stellar environments described in the previous sub-section are likely to be mainly in the form of carbons and silicates. They are ejected into the interstellar medium and may be affected by processes occurring there. The main interstellar agents for modification of the dust properties are irradiation by the mean interstellar radiation field, and modification in shocked hot gas. Cosmic rays may also affect dust properties.

The mean interstellar radiation field is dominated by powerful radiation from the (relatively few) massive hot stars. The intensity of this radiation peaks in the ultraviolet, although photons with energies greater than 13.6 eV are trapped in ionized zones around those stars (see Sections 3.3 and 5.2). What effect can a radiation field with photons of energy up to 13.6 eV (i.e. wavelengths longer than 91.2 nm) have on the likely components of interstellar dust? This radiation is capable of altering the chemical and optical properties of hydrogenated carbons. Since the carbons are formed in regions in which hydrogen is abundant, it is very likely that hydrogen is incorporated in carbon dust during the formation process, so that this dust is perhaps better described as hydrogenated amorphous carbon. The interstellar radiation field is capable of removing much of the hydrogen from hydrogenated amorphous carbon, and according to laboratory experiments, should be able to do this in the interstellar medium in timescales of about a million years. The removal of hydrogen changes the physical nature and chemical structure of the material from one in which the carbon atoms exhibit a valency of four (as usually found in carbon-based polymers) to a valency of three (as found in graphene); the change also changes the optical properties of the material in the visual part of the spectrum, and the process is called *photodarkening*. The process can be reversed by hydrogen insertion reactions, but these only occur at temperatures higher than those normally found in cold interstellar clouds. Evidently, the optical properties of carbon dust grains depend on the environment of the dust, and the interstellar extinction curve should vary from one line of sight to another (as observed), to take account of variations of the local interstellar radiation field on the dust properties, the local temperature, and the atomic hydrogen abundance. Silicates, however, are not so sensitive to the local radiation field.

We shall describe the origins and effects of shock waves in the interstellar medium in detail in Chapter 6. All that we need to know for present purposes is that shock waves deposit energy very abruptly as they pass, heating a narrow region of space. The post-shock temperatures depend on the shock speed, with shock speeds greater than about 50 km s^{-1} raising the post-shock temperature sufficiently to ionize the gas. Even low velocity shocks achieve post-shock temperatures of a thousand K or more, sufficient to drive chemistry through reactions that have a significant activation energy barrier (see Section 6.6). We consider here the effects of high temperature post-shock gas on interstellar dust grains. In hot post-shock gas, atoms and ions striking grains may erode the surface and eject dust material as atoms, ions, or molecules; this process is called *sputtering*. Grain–grain collisions may also occur, because the motion of charged grains (as we'll see in Section 4.4.2) is affected by the magnetic field. These grain–grain collisions cause *shattering* in which some of the material of larger grains is redistributed into a number of smaller fragments. For sufficiently high shock velocities, *catastrophic shattering* may occur. In these collisions, the target grain is completely shattered into fragments. Thus, the passage of grains through shocks is to modify the initial grain size distribution, redistributing grain material from large grains to small. Therefore, the size distribution arising from the formation of dust is not the size distribution of interstellar grains inferred from extinction observations. As we saw in Section 4.2.2, the inferred size distribution for interstellar dust is strongly skewed to small grains; in the low-density interstellar medium, there are typically about 3000 more small (radius ~0.01 micron) grains per unit volume than large (radius ~0.1 micron) grains. This conclusion is consistent with theoretical studies of sputtering and shattering.

Of course, the consequences of sputtering and shattering depend on the chemical and physical nature of the dust. Water ice grains are readily sputtered in shocks with velocity as low as a few km s^{-1}, while grains of other materials are mainly affected by shattering in shocks with much higher velocities. The physical nature of dust, such as the porosity, also affects the shattering in shocks.

Grains of hydrogenated amorphous carbon contain planar structures similar to small pieces of graphene, and erosion in interstellar shocks will release these structures into the interstellar gas to become free-flying molecules. These structures contain hexagons of carbon atoms that make up the structure of graphene, and are called *polycyclic aromatic hydrocarbons* (or PAHs). PAHs may contain several hexagons attached to other radicals. While no identification of a specific PAH molecule in the interstellar medium has yet been made, all PAHs have characteristic infrared spectra with features including those at 3 and 11 microns, and these features are widely observed in the interstellar medium.

One of the many possible fates that may befall dust grains is that they are involved in regions of star and planet formation where they may be incorporated and processed into solid components such as asteroids, comets, and planetesimals. Erosion of such bodies may populate these regions with dust, that is, with interplanetary dust, see Figure 4.5.

FIGURE 4.5 Some interstellar dust may be included in star and planet forming regions, forming solids such as asteroids, comets, and planetesimals. Erosion of such bodies contributes to interplanetary dust. The image shows the erosion of Comet 67P/Churyumov-Gerasimenko. Credit: ESA/Rosetta/NAVCAM, CC BY-SA IGO 3.0

4.4 PHYSICAL PROPERTIES OF GRAINS

4.4.1 Temperature

In Section 4.2.1, we defined the efficiency for absorption of radiation by a grain as Q_{abs}. In clouds of low density, the most important way energy is transferred to the grains is by absorption of photons. Collisions are not so important. A UV flux of 10^{10} photons m^{-2} s^{-1} nm^{-1} in the low-density interstellar medium, with $Q_{abs} = 1$, deposits 10^{-20} J s^{-1} into each unshielded grain, whereas collisions deposit $\lesssim 10^{-26} n_H$ J s^{-1} per grain.

The grains, therefore, readjust their temperatures by re-radiating, but at a lower temperature, T_g, than the stellar radiation temperature. T_g is found from the equation

$$\int F(\lambda)Q_{\mathrm{abs}}(a,\lambda)d\lambda = \int Q_{\mathrm{abs}}(a,\lambda)B(\lambda, T_{\mathrm{g}})\,d\lambda \tag{4.7}$$

in which $F(\lambda)$ is the energy flux of stellar radiation, and B is the Planck function

$$B(\lambda, T_{\mathrm{g}}) = \left(2hc^2/\lambda^5\right)\left(exp\left[hc/k\lambda T_{\mathrm{g}}\right] - 1\right)^{-1} \tag{4.8}$$

The two integrals contribute over different wavelength ranges. The left-hand side calculates energy input into the grain from the radiation field in the visible and UV. The right-hand side calculates the energy emitted, and this range of wavelengths is generally in the infrared. If the grain were a perfect radiator, its temperature would reflect the energy content of space and be equal to the black-body temperature $T_{\mathrm{BB}} \approx 3$ K. At this temperature, however, the peak of the Planck function occurs in the millimetre range, and grains of approximately 10^{-7} m radius cannot readily radiate at these wavelengths. Consequently T_g is higher than T_{BB}. Illustrative grain temperatures for silicate and graphite grains in an unshielded mean interstellar radiation field are calculated to be about 16 and 22 K, respectively. Temperatures inside clouds are, of course, lower. Nevertheless, because interior grains absorb infrared photons emitted by hotter grains at the edges of a cloud, interior grain mean temperatures are unlikely to fall as low as T_{BB}, ~ 3 K. For very small grains, the random arrival of photons may cause considerable variation in grain energy content. The grain temperature may therefore show severe ‘spiking’, in which the temperature may sometimes be far from its mean value. We shall see that grain temperature can be an important parameter in determining mechanisms by which grains may act as catalysts (Section 4.6).

4.4.2 Grain Electric Charge

This is another parameter which is important in connection with the rate at which ions may collide with and stick to dust grains. Ions in the gas, such as C^+ or H^+, will have either enhanced or inhibited collision rates with grains depending on whether the grain is charged negatively or positively. In this section, we briefly discuss the physical effects which control the charge on a dust grain. Obviously, collisions of positive ions and electrons with grains will affect the charge, and therefore alter the rate of such collisions on grains. Let us assume that each such particle that hits a grain actually sticks to it.

Consider a positive ion of mass m_i, and positive charge e approaching a spherical grain of radius a carrying a charge of Z electrons. It follows the path shown in Figure 4.6, on which it is *just* captured by the grain due to electrostatic

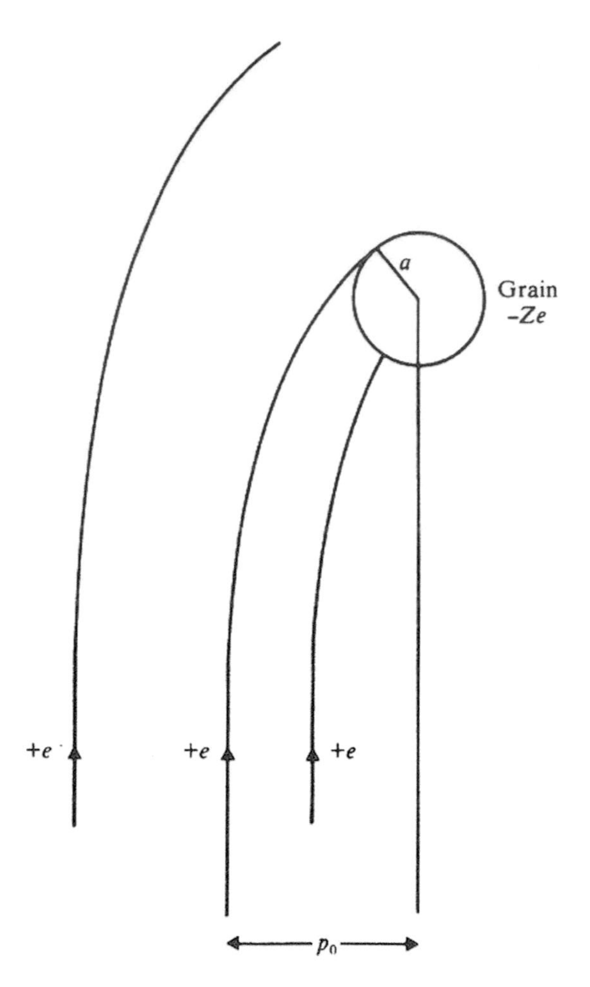

FIGURE 4.6 The capture of a positive ion charge +e by a grain carrying negative charge −Ze. The impact parameter at which a grazing collision occurs is p_0.

attraction. We say that it has an impact parameter p_0. For all impact parameters p that are less than p_0 capture occurs, so that the capture cross-section is πp_0^2. If the velocity of the ion relative to the grain is u_i, when they are well separated, and v, when at the point of capture, then by conservation of angular momentum

$$u_i p_0 = av \tag{4.9}$$

and by conservation of energy

$$\frac{1}{2} m_i u_i^2 = \frac{1}{2} m_i v^2 - \frac{Ze^2}{4\pi\varepsilon_0 a} = \frac{1}{2} m_i \frac{u_i^2 p_0^2}{a^2} - \frac{Ze^2}{4\pi\varepsilon_0 a} \tag{4.10}$$

from which we find the capture cross-section

$$\sigma_i = \pi p_0^2 = \pi a^2 \left(1 + \frac{2Ze^2}{4\pi\varepsilon_0 a m_i u_i^2}\right) \tag{4.11}$$

for ions being captured. Similarly, for electrons the capture cross-section is

$$\sigma_e = \pi a^2 \left(1 - \frac{2Ze^2}{4\pi\varepsilon_0 a m_e u_e^2}\right) \tag{4.12}$$

where m_e is the electron mass and u_e is the electron's velocity when it is far away from the grain. In equilibrium, the rate of collisions of electrons and ions must be equal if no other process plays a part; we should calculate these rates by integrating over Maxwellian distributions in velocity for ions and electrons in equations (4.12) and (4.13). However, we can obtain an approximate measure of the charge by using the most probable velocities $\bar{u}_i$ and $\bar{u}_e$. Equilibrium means that

$$n_e \sigma_e \bar{u}_e = n_i \sigma_i \bar{u}_i. \tag{4.13}$$

If the cloud is overall neutral, then $n_e = n_i$, and from equations (4.11), (4.12), and (4.13)
we find

$$Z = \frac{3kT/2}{e^2/4\pi\varepsilon_0 a} \frac{\bar{u}_e - \bar{u}_i}{\bar{u}_e + \bar{u}_i} \approx \frac{3kT/2}{e^2/4\pi\varepsilon_0 a}. \tag{4.14}$$

At T = 100 K, Z is therefore of the order of 1 for $a = 10^{-7}$ m, while at T = 10 000 K, Z is approximately 100.

There is another effect which may be important in determining the grain charge: photoelectric emission from grains. At visible wavelengths, the efficiency of photoelectric emission, y, is rather small, near 10^{-4}; that is, about 10^4 photons may be absorbed to cause the emission of one electron. For photons in the far UV ($\lambda \approx 100$ nm), efficiencies are probably higher, especially in the case of very small particles ($a \approx 10$ nm) from which the electrons released escape easily. A typical interstellar photon flux near $\lambda = 100$ nm is 10^{10} photons m^{-2} s^{-1} nm^{-1}. We might expect about 10^{11} y electrons to be released per m^2 of grain surface per second in an unshielded region. This may be compared with a collision rate for electrons of $n_e\ v_e$ m^{-2} s^{-1} (if the accumulated charge is not large). We see that if y can be as large as 0.1, then typical low-density clouds in which $n_e \approx 10^5$ m^{-3} have grains whose charge is dominated by the photoelectric effect and are therefore positively charged. If this is so, then collisions of C^+ and H^+ will be partially suppressed, with consequences for catalysis involving these ions.

4.5 DUST GRAINS AS HEATING OR COOLING AGENTS IN INTERSTELLAR GASES

It is possible that dust grains may contribute to the interstellar heating and cooling processes, which we discussed in Chapter 3. Certainly, the photoelectric effect which we mentioned in the previous section would add to the heating of the gas, for each electron released would carry with it several eV of energy. The energy, E, of the emitted electron is given by

$$E = h\nu - W \tag{4.15}$$

where W is the work function of the material (typically about 5 eV) and $h\nu$ is the energy of the UV photon, approximately 11 eV. With the parameters used in Section 4.4.1, we find a heating rate of about $10^{-26} ny$ J m^{-3} s^{-1}, which by comparison with the values in Section 3.3.2 may be important, if y is not too small (cf. Section 4.4.2).

Grains may also act as a sink for energy. If an H atom in a 100 K gas hits and sticks to a grain at temperature 20 K, then an energy equivalent to 80 K has been lost from the gas, and will be present in the grain and ultimately radiated away. However, as we'll see in sub-sections 4.6.1 and 4.6.2, observations require that nearly all hydrogen atoms colliding with dust grains must leave the surfaces combined in hydrogen molecules. This chemistry can be highly exothermic, and the energetic hydrogen molecules produced are a powerful heat source for the interstellar gas. In comparison, the cooling mechanism described above contributes a negligible amount of cooling.

An interesting situation can arise, however, when a dense cloud is heated and becomes optically thick in IR radiation. Then the collection of dust grains radiates like a black-body. The peak wavelength λ_m (m) of a black-body at temperature T_{BB} (K) is given by

$$\lambda_m T_{BB} \approx 3 \times 10^{-3} \text{m K} \tag{4.16}$$

so that if the grains are at, say, 300 K, the emitted radiation is predominantly around 10 μm. Such sources are observed, and are very bright in the infrared. They occur when all the UV radiation from a powerful star is trapped inside a dense cloud of gas and dust. Such a phenomenon – sometimes called a cocoon star – may arise in the process of star formation when a star begins to shine brightly although still inside the cloud which gave birth to it.

4.6 BARE GRAINS AS SITES OF MOLECULE FORMATION

4.6.1 H_2 FORMATION ON BARE GRAIN SURFACES

We suggested in Section 3.5 that the question of how hydrogen molecules form in the interstellar gas might be answered by the contribution made by surface reactions on interstellar grains. It need not be surprising that these reactions have some effect, for we know that in the laboratory the recombination of atoms into molecules occurs readily on the surfaces of vessels containing atomic gas (as well as in three-body collisions in the gas if the number density is high enough). The surfaces bring H atoms together and allow stabilization of the molecule to occur by absorbing some energy from the colliding pair. The major question really is this: can interstellar dust grains be efficient enough in producing molecules so as to compete with the H_2 loss mechanisms, which we described in Chapter 3?

Let us try to answer this question by assuming that a catalytic process for

$$H + H \rightarrow H_2 \tag{4.17}$$

is completely efficient for all H atoms arriving at a grain surface. Then – assuming the best case that H atoms only leave the grain as part of an H_2 molecule – the rate of formation of H_2 is simply determined by the rate of arrival of H atoms at the surfaces of dust grains:

$$\frac{d}{dt} n(H_2) = \frac{1}{2} n(H) \pi a^2 n_g v_H \tag{4.18}$$

for grains of radius a, number density n_g, and where v_H is the most probable velocity of H atoms. Now the quantity $\pi a^2 Q_{ext} n_g$ is related to extinction, and if we assume $Q_{ext} \approx 1$, then the grains causing visual extinction give a value for $\pi a^2 n_g$ of $3 \times 10^{-26} n$ m^{-1}, so that H_2 forms in a gas at ~100 K at a rate

$$\frac{d}{dt} n(H_2) \approx 3 \times 10^{-23} n\, n(H) \quad m^{-3} s^{-1} \tag{4.19}$$

(see Section 3.4.3), where $n = n(H) + 2n(H_2)$. The formation rate will be larger than this because other smaller grains are also present, not contributing significantly to visual extinction, but providing more grain surface area per unit volume.

The mechanism by which H_2 molecules are destroyed has been described in Section 3.3.2. Since this is a process which depends on the absorption of radiation in spectral *lines*, the molecules at the edge of a cloud may shield molecules of the interior by using up photons in the lines causing the destruction of the molecules. The photodissociation rate for H_2 varies strongly with column density, $N(H_2)$ m^{-2}, being approximately 10^{-10} s^{-1} for small $N(H_2)$ ($<10^{18}$ m^{-2}) but being 10^{-14} s^{-1} for large $N(H_2)$ ($\approx 10^{24}$ m^{-2}). In the centre of a cloud of moderate density – say $n = 10^8$ m^{-3} – the formation rate is therefore adequate to convert the bulk of hydrogen to molecular form, in which it is protected against photodissociation by self-shielding. Detailed studies show reasonable harmony between theory and observation. We conclude that grains are potentially capable of contributing significantly to H_2 formation in diffuse clouds.

Let's consider some simple ideas about the surface reaction. To form a hydrogen molecule on a grain surface a minimum requirement is that one H atom is retained at the surface long enough for a second H atom to arrive and locate the first. An H atom in the vicinity of a grain experiences long-range van der Waals forces arising from its interaction with all the atoms of the grain. If the atom is not chemically bound as it approaches close to the surface, these weak forces create a potential well of depth q, where q/k is typically a few hundred K, for the infalling atom. On collision, the atom will generally transfer some energy to the lattice by exciting lattice phonons, and so will remain bound to the surface. It may then move laterally across the surface, or it may become bound to a particular site in the lattice. Lateral motion will be impeded by another energy barrier, typically of energy equivalent to 50 K for physical adsorption on a perfect crystal. It is most unlikely, however, that grains are perfect crystals because of the violent events to which they are subject. There will be sites of dislocation, imperfections, and vacancies, which cause binding sites of energies ranging from weak physical adsorption to strong chemisorption. To ensure that the second H atom arrives on the grain before the first has left, we must require, at least, that the rate of arrival of H atoms on a particular grain exceeds the evaporation rate, that is,

$$n(\mathrm{H})\pi a^2 \bar{v}_{\mathrm{H}} > \nu e^{-q/kT_{\mathrm{g}}}. \tag{4.20}$$

Here, πa^2 represents the cross-sectional area of the grain and $\bar{v}_{\mathrm{H}}$ is a most probable H atom velocity. The frequency of oscillation of the bound H atom perpendicular to the surface is ν inside the well of depth q, and the grain temperature is T_g. We are here assuming that a classical description is adequate, and the exponential gives the fraction of a Maxwellian distribution with energy sufficient to overcome the energy barrier to evaporation, q.

Hence,

$$T_{\mathrm{g}} < \frac{q}{k}\left(\ln \frac{\nu}{n(\mathrm{H})\pi a^2 \bar{v}_{\mathrm{H}}}\right)^{-1}. \tag{4.21}$$

For a physically adsorbed atom (i.e. adsorbed by weak forces, not chemisorbed), this gives $T_g \lesssim 12$ K for numbers given earlier. If $T_g > 12$ K, in this case, then the second H atom arrives too late! From our discussion of temperatures (Section 4.4.1), we see that T_g is unlikely to satisfy this condition. If so, we infer that H atoms must be more strongly bound to the surface for H_2 formation to occur.

Motion across the grain surface affects the rate at which the adsorbed atoms find each other. If they have relaxed to the grain temperature and are bound to lattice sites, then they can travel only by quantum-mechanical penetration of the barrier between the sites. We can describe the motion in a weak binding case in the following way: the atom vibrates laterally with a frequency v' and has a probability P of penetrating the barrier to the neighbouring site a distance a' away. The atom therefore migrates over the surface with a velocity $u = Pa'v'$. Clearly P and v' are dependent on T_g, but a simple calculation in elementary quantum mechanics gives typical values $P \approx 10^{-3}$, for

$a' \approx 3 \times 10^{-10}$ m and $\nu' \approx 3 \times 10^{11}$ s, so that $u \approx 0.1$ m s^{-1}. If the atom is likely to interact with another atom within a distance $l \approx 3 \times 10^{-10}$ m, it will search the whole surface area, $4\pi a^2$, of the grain in a time $4\pi a^2/lu$, which is $\sim 10^{-3}$ s for the above numbers. This time must be short compared with the residence time, $(1/\nu)e^{q/kT_g}$. These simple calculations indicate that an H atom arriving at a surface and being bound at a moderate-energy site so that it is retained long enough for a second atom to arrive will usually meet the second atom. We assume that they then form a molecule. The confirmation of these simple classical ideas must be sought from quantum mechanical studies of atom–surface interactions and especially from realistic laboratory studies.

4.6.2 H_2 Formation on Grains: Fundamental Theory and Laboratory Experiments

Most theoretical studies are made for the graphene structure, since this regular surface of carbon hexagons can be modelled accurately. An incident H atom approaching this surface is found to be initially weakly adsorbed – physisorbed in a van der Waals well – at about 0.3 nm above the surface in a bond of strength up to about 38 meV (or about 440 K equivalent), dependent on the precise binding site. The H atom may also become chemisorbed (in which the electronic structure is modified) but this requires a distortion of the graphene lattice which, in effect, creates an energy barrier of about 0.2 eV to chemisorption. If the chemisorption bond can be established, then its strength is about 0.67 eV. The barrier to lattice distortion would seem to preclude chemisorption at the low temperatures of interstellar clouds. However, further studies show that the barrier to chemisorption can be reduced or removed by the presence of adsorbed H atoms close to the binding site. In fact, theoretical investigations show that the more realistic the model, the more likely that barrier-free chemisorption is predicted to occur. We conclude that H atoms both chemisorbed and physisorbed on carbon grains should be available for reaction with incident H atoms. The theoretical values of these binding energies are confirmed in laboratory experiments.

Reactions of gas phase H atoms with these adsorbed H atoms are predicted by detailed studies to be efficient. Where the release of the H_2 molecule from the surface is prompt, then much of the available energy in the H–H bond (4.5 eV) remains in the resulting H_2 molecule as rovibrational energy. Some of the energy is given to the surface in the process of ejection of the H_2 molecule from the surface, and some appears in the kinetic energy of the molecule. Alternatively, depending on the nature of the reaction site, the newly formed H_2 molecule may remain bound at the site, in which case all the energy released in the reaction is transferred to the grain. This H_2 may be released later, when the solid surface is gently warmed.

These theoretical predictions are confirmed by many remarkable laboratory experiments. These investigations use a variety of materials to represent the surfaces of dust grains, including amorphous carbon and silicates. The experiments are carried out at ultra-high vacuum and at low gas temperature, but the implied gas number density is still very much larger than found in low temperature interstellar clouds. The surfaces are typically held at low temperatures, ~10 K. In some experiments, separate beam lines deliver pure H atoms and pure D atoms to the surface, so that detection of HD molecules is clear evidence that hydrogen surface chemistry has occurred. In another type of experiment, pure H atom gas is used in a single beam line, and the product H_2 that is promptly released is identified by its high vibrational excitation. All these experiments show that H_2 is formed efficiently on the low temperature surfaces selected to represent interstellar grains. The H_2 formed may be retained on the surface and released in later heating, or it may be ejected with high internal and kinetic excitation. The implied efficiency of H_2 formation in surface reactions is consistent with the observational requirement that most of the H atoms in diffuse clouds that strike grain surfaces must leave the surface as part of an H_2 molecule.

Similar experiments suggest that species other than hydrogen can be hydrogenated at the surface of grains. While the hydrides of oxygen and carbon can be formed more efficiently in gas phase reactions (see Section 3.4) than in surface reactions, hydrides of nitrogen do not readily form in this way. Experiments indicate that nitrogen atoms may be hydrogenated efficiently in low temperature surface reactions all the way to ammonia, NH_3. Models suggest that NH, NH_2, and NH_3 abundances in interstellar clouds may arise from surface reactions.

4.6.3 Steady-State H Atom Abundance in Molecular Clouds

Each cosmic ray ionization of H_2 leads to the production of two H atoms, because the H_2^+ ion created by the cosmic ray reacts with another H_2 to form H_3^+, ejecting one H atom, and the recombination of H_3^+ with an electron creates an H_2 molecule, releasing a second H atom. These H atoms recombine to form H_2 molecules in reactions at the surfaces of dust grains, as we have seen in Section 4.6.2 at a rate $k'' \, n \, n(\mathrm{H})$ m^{-3} s^{-1}, where k'' is thought to be about 3×10^{-23} m^3 s^{-1}, and $n = n(\mathrm{H}) + 2n(\mathrm{H_2})$. If these are the only processes forming and removing atomic hydrogen in dark clouds, then in steady state

$$k''nn(\mathrm{H}) = \zeta n(\mathrm{H_2}) \tag{4.22}$$

where ζ is the cosmic ray ionization rate (~ 10^{-17} s^{-1}) so that

$$n(\mathrm{H}) = \frac{\zeta}{k''}\frac{n(\mathrm{H_2})}{n} \tag{4.23}$$

Since the hydrogen is nearly all H_2, $n(\mathrm{H_2}) \simeq \frac{1}{2}n$, and so the hydrogen atom number density is

$$n(\mathrm{H}) \sim 10^5 \mathrm{m}^{-3} \tag{4.24}$$

independent of density. In fact, this is a lower limit, as the cloud is unlikely to be in steady state so it retains more atomic hydrogen from an earlier phase. Thus, in steady state, the fractional abundance of atomic hydrogen in dense clouds is really quite small.

4.6.3 Grains as Sites for Ice Deposition

The infrared spectrum of dust in darker regions of interstellar molecular clouds exhibits a broad (0.3 μm) infrared absorption of amorphous ice at 3.1 μm due to the O–H stretch mode in H_2O (see Figure 4.3). This feature appears in clouds at depths at which the intensity of the external radiation field is typically reduced by more than about one order of magnitude from its mean intensity because of extinction by dust. This suggests that nearer the edge of the cloud, any water molecules formed and retained on the surfaces of grains are removed by photodesorption by ultraviolet photons of the interstellar radiation field. The ultraviolet dissociates H_2O into H and OH, each with excess energy so that desorption may occur. This photodesorption process has an efficiency of about 10^{-3} per UV photon. However, at larger depths, the UV intensity is weaker and the photodesorption of surface water molecules is swamped by H_2O formation, and amorphous ice accretes on the surface of the grains. Water is not abundant in the gas phase, and water ice mantles cannot arise from simple freeze-out of water molecules on to the grains; this would take too long. Evidently, oxygen atoms are converted to water molecules in reactions on the surfaces of grains.

At sufficient depths in molecular clouds, grains that are sufficiently cold accrete 'mantles' of amorphous water ice; however, this ice certainly not pure. As Figure 4.3 shows, infrared signatures are also found of carbon monoxide (CO), carbon dioxide (CO_2), methanol (CH_3OH), ammonia (NH_3), methane (CH_4), and a cyanide-bearing species (OCN^-). The abundances of these species relative to the most abundant species, H_2O, vary widely and especially between dark clouds that are relatively near to bright stars and those clouds that are far from bright stars. This variation suggests that stellar UV radiation can affect the solid-state chemistry that occurs in the ice. The numbers of CO and CO_2 molecules together in the ice may amount to several tens of percent of the number of water molecules, while CH_3OH molecules are typically about 10%, NH_3 and CH_4 molecules are each typically less than 10%, while the cyanide molecules are less than 1% of the number of water molecules.

The variety of molecular species observed in the ice is confirmation of an active surface chemistry, in a process similar to that of H_2 formation (Section 4.6.2). Methane and ammonia form from surface reactions in which carbon and nitrogen atoms arriving on a grain surface from the gas are successively hydrogenated. Methanol is formed from successive hydrogenations of CO, and CO_2 from reactions of CO with OH produced by the photodissociation of H_2O and retained in the ice. The cyanide origin is unclear, but must require surface reactions involving all of C, N, and O atoms.

4.6.4 Ice as Feedstock for Chemical Complexity

As we shall see in Chapter 8, star formation occurs because of gravitational collapse of interstellar clouds, so we expect to find the youngest stars in association with the densest interstellar gas. This idea is supported by observations which show very dense (~ 10^{13} H_2 m^{-3}) and warm (T ~ 300 K) tiny gas cores in which are embedded massive protostars, that is, stars that have yet to ignite their nuclear reactions. These regions are radiating energy obtained from the gravitational collapse, and are commonly called 'Hot Cores'. Cores around solar mass protostars have fairly similar properties and are called 'Warm Cores'. These dense cores (especially one near the centre of the Milky Way known as Sagittarius B2 Large Molecule Heimat)) are rich in 'complex' molecules, that is, molecules with more than just a few atoms; these molecules are almost entirely organic (carbon-based). These molecules with about 10 atoms or so are not 'complex' for chemists, of course, but they are relatively complex when compared to more widespread interstellar molecules such as carbon monoxide or formaldehyde (see Table 3.2). It is possible that these complex organic molecules (or COMs) may be important in the formation of very much larger molecules related to astrobiology. A few examples of molecules detected in active regions of the interstellar medium and considered by astronomers to be 'complex' are listed in Table 4.2.

Gas phase chemical networks (of the kind we discussed in Chapter 3) to form species such as those listed in Table 4.2 can be difficult to design, and often suffer from a lack of gas phase reaction rate data. Some COMs may lack a plausible gas phase route, while some networks seem unable to deliver the required species in adequate amounts. Consequently, production of complex species from an initial material similar to that locked up in interstellar ices has

TABLE 4.2

Some examples of detected astronomically complex organic molecules (COMs).

$HCOOCH_3$ Methyl formate	CH_3OCH_3 Dimethyl ether	$(CH_3)_2CO$ Acetone	CH_3CH_2OCHO Ethyl formate
CH_3COOH Acetic acid	CH_3CH_2OH Ethanol	$(CH_2OH)_2$ Ethylene glycol	CH_3COOCH_3 Methyl acetate
CH_2OHCHO Glycolaldehyde	CH_3CH_2CN Ethyl cyanide	CH_3CHCH_2O Propylene oxide	C_3H_7CN Propyl cyanide
CH_2CHCHO Propenal	CH_3CONH_2 Acetamide	CH_3CH_2CHO Propanal	CH_3OCH_2OH Methoxymethanol

been investigated in a number of laboratory experiments. In these experiments, molecular ices are deposited typically at low temperatures and irradiated with energetic radiation such as UV or X-rays, or with particles such as energetic electrons, or atomic nuclei, to represent processing in the interstellar medium. The product molecules are examined within the ice or after evaporation during gentle warming. In general, a wide range of species similar to those observed in hot and warm cores is readily produced from ices of pure methanol (including dimethyl ether, formic acid, acetic acid, acetaldehyde, ethanol, ethylene glycol, etc.); the energy source seems to be of secondary importance. More complex ices generate an even wider range of products.

Some theoretical studies of chemistry in hot cores assume that molecular species in the ice, like those discussed in Section 4.6.3, are partially photodissociated by the weak local radiation field. These radicals (such as CH_3 from methane and OH from water) should become mobile in the ice as it is warmed by the nearby protostar to about 30 K. The radicals are then able to react and form products (such as CH_3OH, methanol) that evaporate as the ice is warmed further to about 100 K by the nearby protostar. These models are successful in predicting a very wide range of complex species, and the predictions are consistent with the results of laboratory experiments. For example, ices with a composition as found in molecular clouds (H_2O, CO, CO_2, CH_3OH, CH_4, and NH_3) will, under irradiation, develop a population of reactive radicals (including H, OH, CH_3O, CH_2OH, HCO, CH_3, CH_2, CH, NH_2, and NH), which become mobile as the ice is warmed, so that a very wide variety of new products – almost all of which are found in hot cores – will form in simple addition reactions. In a continually warming environment (as exists in a hot core), these new species enter the gas and are able to be detected there.

PROBLEMS

1. Suppose that the interstellar medium contains dust grains with uniform number density 10^{-6} m^{-3}, all with the same radius, 10^{-7} m, and extinction efficiency $Q_{ext} = 0.5$ at wavelength λ_0. Find the extinction in magnitudes at wavelength λ_0 for a star at a distance 1 kpc from the Earth.
2. Derive equation (4.6). What assumption is made concerning n_i? Consider an interstellar cloud in which the number density of CO molecules is 10^6 m^{-3}, and in which the temperature is 10 K. How long would it take for a grain of radius 10^{-7} m to grow to a radius of 2×10^{-7} m by adding a mantle of CO? Assume that the sticking coefficient of CO is unity and that the density of solid CO is 10^3 kg m^{-3}.
3. Estimate a lifetime for interstellar grains in a uniform interstellar gas, H- atom density 10^6 m^{-3}, by sputtering in supernova blast waves, using the information in Section 4.3.2 and assuming a supernova rate for the Galaxy of one every 30 years.
4. Suppose that the heat input into an interstellar grain is entirely by UV absorption. Assume that the UV flux is 10^{10} photons m^{-2} s^{-1} nm^{-1}, that

the bandwidth is 100 nm, and that the mean photon energy is 9 eV. If the grain cools by radiating like a black-body with an efficiency of 0.1%, find the grain temperature.

5. At a point P in an interstellar cloud, $n(\mathrm{H}) = 10^7\ \mathrm{m}^{-3}$ and $n(\mathrm{H_2}) = 4.5 \times 10^7\ \mathrm{m}^{-3}$. Find the photodissociation rate, $\beta(\mathrm{H_2})$, at P. In a plane-parallel slab model, an approximation for $\beta(\mathrm{H_2})$ is $4 \times 10^4 N^{-0.8}\ \mathrm{s}^{-1}$, where N is the column density of $\mathrm{H_2}$ (m^{-2}) towards the cloud edge. Find N at the point P.

5 Radiatively Excited Regions

5.1 INTRODUCTION

Amongst the observationally most spectacular objects in the Galaxy are the large irregular patches of optically visible gas known as diffuse nebulae, or – since they consist mainly of ionized hydrogen – HII regions. They are confined entirely to the plane of the Galaxy and contain large amounts of material, although, in fact, the number densities of gas in them are relatively low (typically about 10^9 hydrogen ions m^{-3}). Their sizes are variable, but linear dimensions of the order of a few parsecs are reasonably representative. Some spectacular examples of HII regions are shown in Figure 5.1.

The optical spectra of these regions are rich in emission lines such as the Balmer lines of atomic hydrogen (corresponding to transitions $n \geq 3 \rightarrow 2$, where n is the principal quantum number. The $3 \rightarrow 2$ transition generates red light of wavelength 656 nm, prominent in Figure 5.1(b)). Also, particularly prominent are lines which originate from transitions between energy levels of ions such as O^+, O^{++}, S^+, N^+, and many others. These lines are known collectively as forbidden lines since they arise from magnetic dipole or electric quadrupole transitions (unlike the much faster allowed electric dipole transitions which produce the hydrogen lines).

Diffuse nebulae also emit in other regions of the electromagnetic spectrum, in particular in the radio region (see Section 5.4). In that part of the spectrum, they produce both a continuous spectrum and spectral lines. The continuum ('bremsstrahlung' or 'free–free' radiation) originates when electrons accelerate in unbound orbits; the lines originate from transitions between very high-energy levels of hydrogen and helium.

The temperature of the gas in HII regions is quite high. Observationally estimated gas temperatures lie roughly in the range 5×10^3–10^4 K, with rather little variation from one nebula to another or within a given object. The source of excitation of this gas is very obvious since diffuse nebulae are always found associated with stars of early spectral type (O–B stars). The effective temperatures of these stars are high (2×10^4–6×10^4 K). They therefore emit a substantial fraction of their radiation output in ultraviolet (UV) photons with energies which are high enough to cause photoionization of elements such as hydrogen and oxygen. The kinetic energy given to the ejected electrons by photoionization is the source of heat for the nebulae.

Diffuse nebulae are ready-made laboratories for studying the interaction of energetic photons with a low-density gas. Although their main properties are

FIGURE 5.1 Images of HII regions. (a) The Orion nebula, one of the brightest nebulae in the night sky and the closest region of massive star formation to Earth. (b) An external galaxy, M51, shows strings of HII regions (in red) along the spiral arms. (Credit: (a) NASA, ESA, M. Robberto (STScI/ESA) and the Hubble Space Telescope Orion Treasury Project Team. (b) NASA, ESA, S. Beckwith (STScI), and the Hubble Heritage Team (STScI/AURA).)

well understood, the physics of these objects provides an excellent basis for the investigation of many important problems in astrophysics.

5.2 NEBULAE OF PURE HYDROGEN

5.2.1 The Structure of Pure Hydrogen Nebulae

Although we saw in Chapter 3 that interstellar gas may contain molecules, we will ignore this fact when discussing the effects of radiation on the gas. Partly this is because the interaction of radiation with molecules is inherently more complex than with atoms. But, more importantly, we shall see that radiatively excited regions contain mainly ions and electrons and a few atoms. Molecules cannot exist with appreciable abundances within these intensely UV-radiated regions. The differences introduced by considering the effects of the presence of molecules in the initially non-ionized gas are not sufficient to justify the extra complexities here.

As we might expect, many – but not all – of the most important features of radiatively excited regions arise from the presence of hydrogen, the most abundant element. It is, therefore, instructive to consider first the physics of radiatively excited regions of pure hydrogen. We treat the nebula as static, though the discussions in Chapter 7 show that this is not the case. However, what is important is that the timescales of dynamic effects are much longer than those associated with atomic and molecular processes. Consequently, as far as the latter processes are concerned, nebulae are indeed static.

5.2.2 Basic Physical Processes

Photons of energies $h\nu \geq I_H$ (the ionization potential of hydrogen) eject electrons from the atoms. The energy excess $(h\nu - I_H)$ goes directly into kinetic energy of the detached electron. The protons are essentially unaffected by this process, although momentum conservation demands that they experience a small recoil. The strong Coulomb interaction between protons and electrons produces the inverse process of recombination. Schematically, we have the reversible reaction

$$\mathrm{H} + h\nu \rightleftharpoons \mathrm{p} + \mathrm{e} \tag{5.1}$$

The energy of the photon produced on recombination is given by the sum of two contributions: these are the kinetic energy of the recombining electron and the binding energy of the electron in the energy level into which it recombines. This binding energy is equal to I_H/n^2, where n is the principal quantum number of the level into which recombination occurs. Following recombination, further photons are produced as the electron cascades downwards through the energy levels, ultimately to the ground level. This process produces lines such as the Balmer and Lyman series of hydrogen. In equilibrium, there is a balance between the backward and forward rates of reaction (5.1). This is called the condition of ionization balance; it determines the degree of ionization of the hydrogen.

5.2.3 The Particle Temperatures

A glance at reaction (5.1) shows that in equilibrium the gas contains three distinct species of particle, namely electrons, protons, and neutral atoms. Energy is being continuously fed into the gas by photoionization of the atoms. This energy is injected in the form of electron kinetic energy. The velocity (or energy) distribution of the ejected electrons will reflect the energy distribution of the ionizing photons. We must therefore consider two important questions. Firstly, do all particles present have Maxwellian velocity distributions, that is, can we ascribe a temperature, in the usual kinetic sense, to the gas particles? Further, if we can, are these temperatures the same or different for the various species of particles?

Fortunately, the answer to the first question is yes, and to the second is that a single temperature suffices. The reason is that the collisional processes which must occur to thermalize particle velocity distributions and to transfer energy from one species of particle to another (e.g. from hot injected electrons to other electrons) take place on such short timescales that they can be regarded as more or less instantaneous. There is a definite hierarchy of processes, which is fixed in sequence according to the timescale associated with each process. We can briefly summarize it as follows. The hot ejected electrons share their energy with other electrons. Electrons then transfer energy to protons until equipartition is reached (note that this process takes a large number of collisions because of the large mass difference between the particles). Finally, energy is transferred to neutral atoms by proton collision until they too come to equipartition.

Note that here we are talking about a redistribution of energy by elastic collisions. The actual value of the temperature is affected by slower inelastic processes which can remove energy from the electron gas (Section 5.3.2). Of course, the remarks about equipartition remain valid if we remove energy from the gas by such inelastic processes (as opposed to injecting it via hot electrons). Redistribution of energy is always so fast that alteration of the energy content of the electron gas is equivalent to alteration of the energy content of the gas as a whole. This also has the important advantage that we can forget what is happening to the gas temperature on microscopic scales when we consider the dynamical behaviour of gases (Chapter 6).

It is conventional to refer to the gas temperature as the electron temperature T_e. As noted in Section 5.1, typical values of T_e lie between 5×10^3 and 10^4 K. We will discuss the reasons for the occurrence of these characteristic values later.

5.2.4 The Recombination of Protons and Electrons

When considering the ionization balance, we will compare recombination and ionization rates per unit volume of gas. Recombination may occur into any level of principal quantum number n. The recombination rate per unit volume is obviously proportional to the product of the electron and proton number densities (n_e and n_p, respectively). In a pure hydrogen nebula, $n_e = n_p$. The recombination rate depends also on the electron temperature, for two reasons. Firstly, the probability of recombination of a given electron into a given energy level depends on the electron's energy. Secondly, the rate at which a given proton encounters electrons depends on the electron velocity distribution, that is, on the electron temperature. The volume recombination rate into level n can be written as

$$\dot{\mathcal{N}}_n = n_e n_p \beta_n(T_e) \equiv n_e^2 \beta_n(T_e) \ \mathrm{m^{-3} s^{-1}} \tag{5.2}$$

$\beta_n(T_e)$ is called the recombination coefficient into level n. For the purposes of the ionization balance, we need the total rate of recombination per unit volume. This is found by summing equation (5.2) over all available energy levels. However, we must be careful about which levels we include in this summation. Suppose an electron of kinetic energy E ($\approx kT_e$ on average – see Section 5.2.8) recombines into level n. The photon produced by this recombination has energy

$$hv = I_H/n^2 + E \tag{5.3}$$

If $T_e \approx 10^4$ K, $E \approx 1.4 \times 10^{-19}$ J. Equation (5.3) can be written in the approximate form

$$h\nu \approx I_{\rm H}(1/n^2 + 0.07) \tag{5.4}$$

As we will see in Section 5.2.5, ionization can be assumed to occur only from the ground ($n = 1$) level. Thus, only recombination directly into the $n = 1$ level can produce photons which can cause further ionization. These recombinations increase the effective ionizing radiation field incident on the gas and these extra photons are usually called the 'diffuse' radiation field. We take into account these extra ionizing photons by assuming that they cause ionization of a neutral atom in the very same region of the gas in which they are created. This is called the 'on-the-spot' assumption. Ground-state recombinations are assumed to be balanced by an immediate ionization. Thus, when we consider the ionization balance between recombination and ionization by photons from the star, we do not include recombinations directly to the ground level. We therefore need the total recombination rate per unit volume, $\dot{\mathcal{N}}_R$, given by

$$\dot{\mathcal{N}}_{\rm R} = \sum_{n=2}^{\infty} \dot{\mathcal{N}}_n \equiv n_{\rm e}^2 \beta_2(T_{\rm e}) \ {\rm m^{-3}s^{-1}}. \tag{5.5}$$

An excellent numerical approximation for the total recombination coefficient, $\beta_2(T_{\rm e})$, is

$$\beta_2(T_{\rm e}) = 2 \times 10^{-16} T_{\rm e}^{-3/4} \ {\rm m^3 s^{-1}}. \tag{5.6}$$

5.2.5 The Ionization of Hydrogen

Hydrogen atoms in any energy level can undergo photoionization, provided photons of sufficient energy are available. As we have seen, atoms can be formed in any excited level by the process of recombination. But the lifetimes of hydrogen atoms in their excited states are very short, since transitions between the energy levels occur by electric dipole transitions which are very fast. Typically, the lifetime of an excited level is measured in fractions of a microsecond. Because of this, the probability of ionizing a hydrogen atom in an excited level is completely negligible. We thus have the great simplification that all the hydrogen atoms can be assumed to be in the ground level for the purpose of calculating the ionization rate. (This also means that the absorption of lines such as the Balmer lines can be neglected.)

To make this calculation, consider a small element of unit volume and unit area face, situated at a distance r from the star (Figure 5.2(a)). Let the number of stellar photons with energies $h\nu \geq I_{\rm H}$ crossing unit area in unit time at r be J m^{-2} s^{-1}. The ionization rate is given by

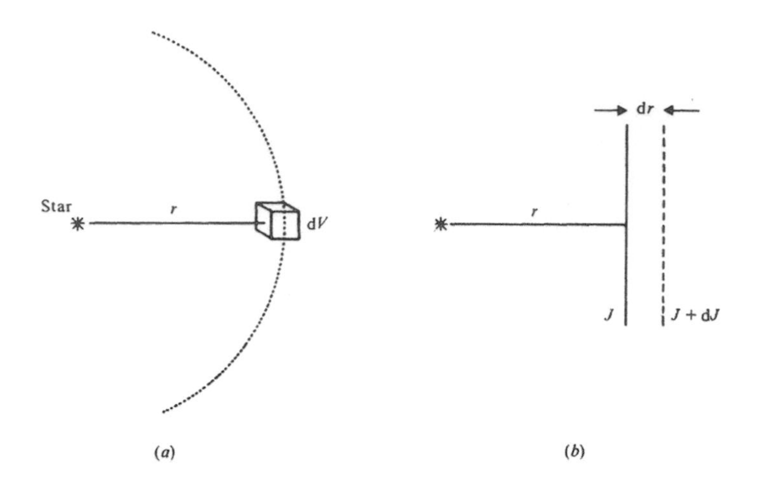

FIGURE 5.2 The geometry for absorption of UV radiation in a nebula.

$$\dot{\mathcal{N}_{\rm I}} = \alpha_0 n_{\rm H} J \ {\rm m^{-3} s^{-1}}. \tag{5.7}$$

In this equation, n_H is the number density of neutral hydrogen atoms and the quantity α_0, which has the dimensions of area, is the photoionization cross-section. It is the effective area for ionization presented to a photon by an atom in the ground level. Strictly speaking, the ionization cross-section depends on the energy of the incident photon and decreases with increasing photon energy. For most purposes, though, it is a convenient and reasonably accurate simplification to take it as a constant equal to the value it has for a photon of energy $h\nu = I_H$. The numerical value of the cross-section is $\alpha_0 = 6.8 \times 10^{-22}$ m^2.

We introduce x, the degree of ionization of hydrogen, by the relationship

$$n_e = xn \tag{5.8}$$

where n is the number density of hydrogen nuclei (i.e. the sum of the proton and neutral atom densities). Thus,

$$n_H = (1 - x)n \tag{5.9}$$

(obviously $\leq x \leq 1$). The condition of ionization balance is

$$\dot{\mathcal{N}_{\rm R}} = \dot{\mathcal{N}_{\rm I}}$$

which, using equations (5.5) and (5.7) becomes

$$x^2 n^2 \beta_2(T_e) = \alpha_0 (1 - x) nJ. \tag{5.10}$$

Rearranging equation (5.10), we find

$$\frac{x^2}{1-x} = \frac{J}{n} \frac{\alpha_0}{\beta_2(T_e)}. \tag{5.11}$$

To estimate J at a typical point in a nebula, assume that J is determined simply by geometrical dilution of the stellar radiation field (i.e. dilution by an inverse square law). As we will see in Section 5.2.7, this is a good approximation except very near the edge of the nebula. Hence,

$$J = S_*/4\pi r^2 \quad \mathrm{m^{-2}s^{-1}}. \tag{5.12}$$

where S_* is the rate at which the star emits photons which can cause ionization (the units of S_* are $\mathrm{s^{-1}}$).

It is instructive to insert representative parameters for diffuse nebulae into the ionization balance equation and examine the consequences. We take $n = 10^8\ \mathrm{m^{-3}}$ and $r = 1$ pc together with a value for S_* of $10^{49}\ \mathrm{s^{-1}}$ (corresponding to a star of main sequence spectral type of about O6.5). Equation (5.12) gives $J \approx 8.4 \times 10^{14}\ \mathrm{m^{-2}\ s^{-1}}$ and the equation of ionization balance (5.11) (with $T_e = 10^4$ K) becomes

$$x^2/(1-x) \approx 2.9 \times 10^4. \tag{5.13}$$

The approximate solution of this equation is $1 - x \approx 3.4 \times 10^{-5}$. We have arrived at a most important and quite general conclusion. The gas in a radiatively excited region is almost fully ionized. Therefore in writing the ionization balance equation, we can put $x \approx 1$ (but not, of course, $1 - x = 0$).

Occasionally, there are circumstances met in astronomy where it is not immediately obvious whether gas, whose temperature is observed to be about 10^4 K, is photoionized or heated by, for example, a shock wave (Chapter 6). If, somehow, we can estimate x independently, our discussion above indicates that a value of x of about unity is a strong argument for photoionization heating.

5.2.6 The Sizes of Ionized Regions

A star cannot ionize an indefinitely large amount of surrounding material. Because recombination occurs continuously within the gas, photons are being continuously absorbed. The volume of gas which the star can ionize is limited to that volume in which the total recombination rate is just equal to the rate at which the star emits ionizing photons. Thus, if this region has radius R_s, the condition of ionization balance for the region as a whole is

$$\frac{4}{3}\pi R_s^3 (xn)^2 \beta_2 = S_*. \tag{5.14}$$

Remembering that $x \approx 1$ (and hence $n_e \approx n$), the radius of the ionized region, R_s, can be found from equation (5.14) and is

$$R_s = \left(\frac{3}{4\pi}\frac{S_*}{n^2\beta_2}\right)^{1/3}. \tag{5.15}$$

This radius is called the Strömgren radius, and the sphere of ionized gas is called a Strömgren sphere. Using the representative values $S_* = 10^{49}$ s^{-1} and $T_e = 10^4$ K as before, $R_s \approx 7 \times 10^5 n^{-2/3}$ pc. If, for example, $n = 10^8$ m^{-3}, $R_s \approx 3$ pc, hence justifying the choice of $r = 1$ pc in the previous section as being representative of a point within a diffuse nebula.

5.2.7 How Sharp-Edged Are Nebulae?

A rather important – but as yet unjustified – assumption implicit in our derivation of R_s is that J must decrease towards the edge of the region, ultimately becoming zero. Hence, x must also eventually decrease to zero. But we derived R_s by putting $x \approx 1$ everywhere. In order that our estimate for R_s be reasonable, x should depart appreciably from unity only when r is close to R_s. Our derivation is therefore based on the assumption that the transition from almost fully ionized to neutral gas takes place over a distance much smaller than R_s. The very large reduction in J at the edge of the nebula is due to absorption and not to geometrical effects. As r increases, J initially decreases due to the geometrical dilution factor. This increases n_H and the increased absorption further decreases J, so that n_H then increases, causing a decrease in J, and so on. It is easiest to see how absorption cuts down J by neglecting the geometrical dilution. We therefore consider these effects using plane parallel geometry (Figure 5.2(b)), which removes the inverse-square variation.

Let the flux of ionizing photons at distance r from the star be again J. The flux at distance $r + \mathrm{d}r$ is $J + \mathrm{d}J$ and any change is now entirely due to absorption between r and $r + \mathrm{d}r$. Then

$$J + \mathrm{d}J = J - \alpha_0 n_H J \mathrm{d}r. \tag{5.16}$$

By expressing n_H in terms of x and n, equation (5.16) can be written as

$$\frac{\mathrm{d}J}{\mathrm{d}r} = -\alpha_0 n(1-x)J. \tag{5.17}$$

Now x is everywhere given by the equation of ionization balance (5.11). We can use this equation to express J in terms of x. We also can differentiate it to express $\mathrm{d}J/\mathrm{d}r$ in terms of $\mathrm{d}x/\mathrm{d}r$. Hence J can be eliminated from equation (5.17) and we obtain the following equation for the spatial variation of x:

$$\frac{\mathrm{d}x}{\mathrm{d}r} = -\alpha_0 n\frac{x(1-x)^2}{2-x}. \tag{5.18}$$

Let us introduce a dimensionless unit of length, λ, defined by

$$\mathrm{d}\lambda = \alpha_0 n \mathrm{d}r. \tag{5.19}$$

Note that $(\alpha_0 n)^{-1}$ is approximately the mean free path of an ionizing photon in neutral hydrogen of density n. Thus λ is a length measured in units of this mean free path. Equation (5.18) then becomes

$$\frac{\mathrm{d}x}{\mathrm{d}\lambda} = -\frac{x(1-x)^2}{2-x} \tag{5.20}$$

and we can easily integrate this to obtain

$$2\ln[x/(1-x)] + 1/(1-x) = A - \lambda, \tag{5.21}$$

where A is a constant of integration. If, for convenience, we choose $\lambda = 0$ at $x = \frac{1}{2}$, then $A = 2$ and equation (5.21) becomes

$$\lambda = 2 - 2\ln[x/(1-x)] - 1/(1-x). \tag{5.22}$$

In order to examine the variation of J with x, define

$$J_{1/2} = \beta_2 n / 2\alpha_0. \tag{5.23}$$

We see from equation (5.11) that $J_{1/2}$ is the photon flux when $x = \frac{1}{2}$. The photon flux at a point where the degree of ionization is x is

$$J_x = \left[2x^2/(1-x)\right] J_{1/2}. \tag{5.24}$$

Table 5.1 gives values of λ and $J_x/J_{1/2}$ for a series of values of x derived from equations (5.22) and (5.24).

We see clearly from Table 5.1 that the degree of ionization and (particularly) the photon flux decrease sharply over a distance $\Delta \approx 10(\alpha_0 n)^{-1}$, that is, over a distance of a few times the mean free path of a photon in the neutral

TABLE 5.1

Variation of ionization and photon flux

x	λ	$J_x/J_{1/2}$
0.9	−12.4	16.2
0.7	−3.0	3.3
0.5	0.0	1
0.3	2.2	0.26
0.1	5.3	0.022

gas. Using the value of α_0 given previously, $\Delta \approx 4.8 \times 10^5 n^{-1}$ pc. If we compare this with the typical value of R_s given in Section 5.2.6, $\Delta / R_s \approx 0.7 n^{-1/3}$. Hence $\Delta \ll R_s$ (since $n \gg 1$) and nebulae are indeed very sharp-edged. Finally, note that if we arrange for J at the star to be high (and thus $1 - x$ very small), it can easily be demonstrated that appreciable changes in J due to absorption take place fairly near the edge of the nebula. This justifies our use of the purely geometrically diluted flux in Section 5.2.5.

Remember that J is the flux of *ionizing* photons. The discussion above shows that the ionizing photons (those with energies $\geq I_H$) are retained within the Strömgren sphere. However, photons with energies less than I_H escape freely into the interstellar medium. This has important consequences for the mean UV and optical interstellar radiation field. The UV radiation field is dominated by radiation from the very powerful O and B stars, even though these stars are relatively few in number. The implication is that the interstellar UV radiation field (outside HII regions) contains only photons that are *incapable* of ionizing hydrogen atoms. Therefore, the mean interstellar radiation field, outside Strömgren spheres, exists for wavelengths longer than 91.2 nm. Nevertheless, the mean interstellar radiation field is capable of ionizing many atomic and molecular species and of photodissociating many molecular species. As we have seen in Chapter 3, this truncated interstellar radiation field is capable of driving a complex chemistry.

5.2.8 The Temperature of a Pure Hydrogen Nebula

The process of photoionization feeds energy continuously into the gas. If there were no means of losing this energy, the gas temperature would increase indefinitely. But we have already met one way in which energy is removed from the gas. The kinetic energy of a recombined electron is converted into radiant energy. This energy cannot be returned to kinetic energy (apart from the energy of photons produced by ground state recombinations) and thus the gas as a whole is cooled. The equilibrium gas temperature is determined by the balance between the heating and cooling rates.

Suppose that, on average, energy $\mathcal{Q}$ J is injected into the gas in each photoionization. The energy input rate per unit volume, $\mathcal{G}$, is thus

$$\mathcal{G} = \dot{\mathcal{N}}_I \mathcal{Q} = \dot{\mathcal{N}}_R \mathcal{Q} \text{ J m}^{-3}\text{s}^{-1}. \tag{5.25}$$

On average, at each recombination, energy kT_e is removed from the gas. (The reason why this is not $\frac{3}{2}kT_e$ as might be anticipated is that the recombination coefficient decreases with increasing electron energy; low-energy electrons therefore preferentially recombine.) The energy loss rate per unit volume is just

$$\mathcal{L} = (kT_e)\dot{\mathcal{N}}_R \text{ J m}^{-3}\text{s}^{-1}. \tag{5.26}$$

In equilibrium

$$\mathcal{L} = \mathcal{G}. \tag{5.27}$$

Hence

$$T_e = \mathcal{Q}/k. \tag{5.28}$$

$\mathcal{Q}$ depends on the radiation field of the star and also (because of transfer effects involving the frequency dependence of the ionization cross-section) on the distance of the particular gas element from the star. If the star radiates as a black-body at a temperature T_* (which is not too unrealistic), we state, without proof (see Problem 3 at the end of this chapter), then the average energy injected per photoionization is given approximately by

$$\mathcal{Q} \approx kT_*. \tag{5.29}$$

We therefore predict from equations (5.28) and (5.29) that

$$T_e \approx T_*. \tag{5.30}$$

Typical values of T_* lie in the range 3×10^4–6×10^4 K and thus values of our estimated T_e lie in the same range. Since our predicted temperatures are much higher than observed, the inference is that we have not included important cooling processes.

In a pure hydrogen nebula, there are three additional cooling processes which, in principle, ought to be taken into account. Thermal plasma emits bremsstrahlung (Section 5.4.1), which is observed at radio frequencies (although actually produced over a much wider frequency range). Electrons can further lose energy by collisionally exciting neutral hydrogen atoms and also by ionizing them. However, even when these processes are included, gas temperatures of about 2×10^4 K are predicted, and this is still far too high. To solve this problem, we must relax our assumption that nebulae are composed only of hydrogen.

5.3 NEBULAE CONTAINING HEAVY ELEMENTS

It was stated in the introduction (Section 5.1) that the spectra of nebulae contain lines emitted by ions of elements other than hydrogen. If we were to measure the total radiant energy emitted by a nebula in the hydrogen lines and the energy emitted from the other elements, we would, in fact, find that more energy was contained in the latter. This is perhaps at first glance rather surprising, since the elements whose ions are responsible for the lines are much less abundant than hydrogen. However, the process of excitation of these lines is extremely efficient and more than offsets the abundance difference.

5.3.1 Forbidden Lines

For simplicity, we will confine our attention to one specific element, namely oxygen. This is about the most important single coolant, and the lines its various ions radiate are particularly useful for investigating the physical structure of nebulae. The first ionization potential of oxygen is practically identical to that of hydrogen. Hence, whenever hydrogen is found to be ionized we also expect to find ionized oxygen. Unlike hydrogen, however, oxygen can be ionized more than once. The second ionization potential of oxygen has a value of 35.1 eV. Although this is much higher than the ionization potential of hydrogen (and the first ionization potential of oxygen), some of the hotter stars associated with nebulae emit enough energetic photons to ionize a significant fraction of the oxygen to the second stage of ionization. Thus, oxygen exists in the forms O^+ and O^{++} within the ionized region.

The energy-level diagrams for the ground states of these two ions are shown in Figure 5.3(a) and (b). (Higher energy states are never excited because the gas is too cool, so their details are unimportant.) Both ions contain energy levels lying a few eV above the ground level. Transitions between them produce lines in the visible region of the spectrum. In addition, the lowest energy level of O^{++} contains fine structure with energy differences between levels of about 0.01 eV. Transitions between these fine-structure levels produce infrared lines. However, all these transitions are strictly forbidden in terms of the quantum mechanical selection rules for the emission of electric dipole radiation. The transitions which take place are the much slower

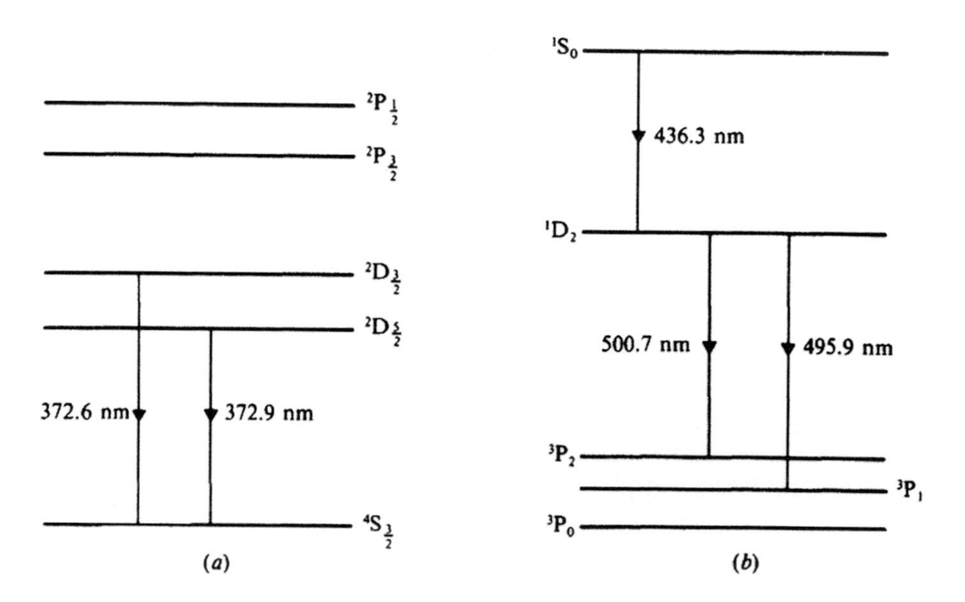

FIGURE 5.3 Energy levels in the ground states of (a) O^+ and (b) O^{++}.

magnetic dipole or electric quadrupole transitions. The transition probabilities are low (typically about 1 s^{-1}) compared with those for electric dipole transitions (typically about 10^8 s^{-1}), and these lines are known as forbidden lines. The levels from which they originate are called metastable levels because of these low transition probabilities.

The identification of observed lines from transitions such as those in Figure 5.3 posed many problems for astronomers in the early part of the 20th century. The reason was that such lines are not reproducible under laboratory conditions. An ion in one of the excited levels can decay by a radiative transition and produce a photon. However, there is a competing process which can suppress the radiation. The excited level can be de-excited to a lower level by electron collision. The transition energy increases the kinetic energy of the colliding electron, that is, the collision is super-elastic. Under laboratory conditions, this latter process dominates and photon emission is drastically reduced. On the other hand, the gas in nebulae is so tenuous that radiative decay generally dominates and photons are readily emitted. Note finally that since the transition probability is low, the corresponding probability of absorption of a photon is also low and these lines always freely escape from the nebulae.

5.3.2 The Excitation of Forbidden Lines

In order that the forbidden lines can be formed, the ions must, of course, be excited to the upper levels of the ground state. In the case of the hydrogen line spectrum, we saw that the mechanism for populating the higher levels was the recombination of a free electron with a proton. In the case of forbidden lines, however, recombination plays a completely negligible role. The basic excitation mechanism is the impact of an electron on an ion on the ground level. Excitation is achieved at the expense of the electron kinetic energy and obviously this process cools the gas. Since the collision frequencies are low compared with the rate at which energy is redistributed amongst the gas particles, the assumption of a unique temperature to describe the various particle energy distributions is always valid.

We shall consider this cooling process only in the low-density limit where we do not have to take into account the possibility of collisional de-excitation from the upper levels. In this limit, every collisional excitation is followed by the emission of a photon. Thus, to calculate the rate of emission of photons per unit volume by transitions from a particular level, we need to calculate the excitation rate to this level per unit volume. A further simplification can be made. In the low-density limit, the timescale for which an ion remains in an excited level is much less than the timescale for which an ion remains in the lower level. Hence it is an excellent approximation to assume that all the ions are present in the lowest energy level and we must know only the number of ions present in a given volume. We need not then consider their distribution amongst the various levels.

Suppose we designate the ground level of an ion (I, say) by i and the upper level by j. Let the total number density of ions be n_{I} m^{-3} (which also equals the number density of ions in level i). The rate of collisional excitation, $\dot{\mathcal{N}}_{ij}$, from level i to level j per unit volume may be written as

$$\dot{\mathcal{N}}_{ij} = n_e n_I C_{ij}(T_e) \text{ m}^{-3}\text{s}^{-1}. \tag{5.31}$$

$C_{ij}(T_e)$ is a collision rate coefficient which depends on the temperature of the colliding electrons (for reasons analogous to those given for the recombination coefficient in Section 5.2.4). Since each excitation is followed by emission of a photon of energy ψ_{ij}, say, the rate of loss of energy per unit volume by this process is

$$\mathcal{L}_{ij} = \dot{\mathcal{N}}_{ij}\psi_{ij} = n_e n_I C_{ij}(T_e)\psi_{ij} \text{ J m}^{-3}\text{s}^{-1}. \tag{5.32}$$

$C_{ij}(T_e)$ has the form

$$C_{ij}(T_e) = \left(A_{ij}/T^{1/2}\right) \exp\left(-\psi_{ij}/kT_e\right) \text{ m}^3\text{s}^{-1}. \tag{5.33}$$

A_{ij} is a constant which depends on the particular ion and transition considered. The exponential factor arises from integration over the Maxwellian velocity distribution of the electrons allowing for the fact that only particles of kinetic energy $\geq \psi_{ij}$ can cause excitation.

The number density, n_I, depends on the proportion of an element in the form of the particular ion considered. If a fraction y_I of a given element z is in the form of ion I, then $n_I = y_I n_z$, where n_z is the number density of nuclei of element z. Further, we can put $n_z = a_z n$, where a_z ($\ll 1$) is the abundance of the element relative to hydrogen. In the case of oxygen, $a_z = 6 \times 10^{-4}$ is reasonably representative. If we take this value, then, as an example, we find the cooling rate due to the collisional excitation of the $^2D_{5/2}$ and $^2D_{3/2}$ levels of O^+ (considered together) is given by

$$\mathcal{L}_{O^+} \approx 1.1 \times 10^{-33} y_{O^+} \left(n^2/T_e^{1/2}\right) \exp\left(-3.89 \times 10^4/T_e\right) \text{ J m}^{-3}\text{s}^{-1}. \tag{5.34}$$

In equation (5.34), we have used the fact that $n_e \approx n$ within a nebula. Note that at temperatures typical of nebulae (5×10^3–10^4 K), the exponential factor depends sensitively on the temperature T_e.

Now let us estimate the temperature of the gas if we include forbidden-line cooling due to the O^+ lines discussed earlier. We will take y_{O^+} equal to unity (i.e. assume all the oxygen present is the singly ionized form). The equation of energy balance is $\mathcal{L}_{O^+} = \mathcal{G}$. Since (by equation (5.25)

$$\mathcal{G} = \dot{\mathcal{N}}_R \mathcal{Q} = n^2\beta_2(T_e)kT_*, \tag{5.35}$$

we find, using the form of $\beta_2(T_e)$ given in equation (5.6), that the equilibrium temperature is given by the solution of

$$T_e^{1/4} \exp(-3.89 \times 10^4/T_e) = 2.5 \times 10^{-6} T_*. \tag{5.36}$$

Calculated values are given in Table 5.2.

TABLE 5.2

T_*	Calculated equilibrium temperature, T_e (K)
2×10^4	7450
4×10^4	8500
6×10^4	9300

These derived temperatures are in general agreement with observed temperatures, and the agreement remains even if other ions are included in the cooling rate. Quite obviously, forbidden-line cooling dominates over all other cooling mechanisms. Note in Table 5.2, the very small variation in T_e is associated with a large variation of T_*. This result is due to a combination of two effects. Suppose, for example, we try to increase T_e by increasing T_*. Since the recombination rate has an inverse dependence on T_e, the recombination rate will decrease and so will the density of neutral hydrogen. This will decrease the heating rate and counteract the effect of an increased Q. Further, at the temperatures considered, an increase in T_e increases the net cooling rate, which further counteracts the effect of the increased T_*. Exactly the converse happens if T_* is decreased. The gas temperature in nebulae is controlled by a fairly sensitive thermostat, resulting in the small temperature range observed.

5.3.3 The Distribution of Ions of Heavy Elements in Nebulae

We have already noted that oxygen can appear as either O^+ or O^{++} within nebulae. Obviously, other atoms exist in more than one stage of ionization. It is important to know how these various stages of ionization are distributed throughout the ionized region. For example, as is evident from the form of equation (5.34), the cooling rate due to transitions within a particular ion depends on the proportion of the element present in that stage of ionization. Further, when we try to build up a model for the structure of a nebula from the radiation emitted by different ions, it is important to know whether the ions exist in the same or different spatial regions of the nebula.

An immediate observational indication that ions of different parent ionization potential may exist in different regions of a nebula comes from direct photographs. Radiation from a particular ion can be isolated by use of a filter which transmits light in a narrow band of wavelengths only. If, for example, a given nebula is photographed in the 372.7 nm doublet lines of O^+ and the resulting photograph compared with one taken in, say, the 500.7 nm line of O^{++}, any difference in spatial distribution of the two ions is immediately apparent.

Very roughly speaking, such a comparison would show that the O^+ emission extends rather farther from the star than the O^{++} emission but with considerable overlap. As a generalization we can say that ions of the highest parent ionization potential are found nearest to the star. This phenomenon is called *ionization stratification* and is the result of the way in which photons of different energies are differentially absorbed as they propagate outwards from the star.

In Section 5.2.6, we discussed the sizes of nebulae on the assumption that hydrogen was entirely responsible for the absorption of the stellar UV photons. If we include all other elements, our conclusions are affected quite negligibly because of their low abundances. However, from the point of view of the energy distribution of the photons, it is necessary to consider the effects of an element which so far we have not mentioned, helium.

Helium is much more abundant than the other heavy elements. Typically, about 10% (by number) of interstellar atoms are helium atoms. The first ionization potential of helium is 24.6 eV, and helium can be present in either the neutral (He^0) or singly ionized (He^+) forms in regions in which the hydrogen is fully photoionized. (Ionization of He^+ to He^{++} requires photons of energy $h\nu \geq 54.4$ eV, which are so few in number for the stars exciting diffuse nebulae that it is completely unimportant.) In Table 5.3 we list a few of the more common heavy elements and their first and second ionization potentials measured in units of 24.6 eV.

It is clear from Table 5.3 that if photons energetic enough to ionize He^0 to He^+ are not available at some point in the nebula then O^{++}, N^{++}, and so on cannot be produced either. Conversely, if these photons are available, we would expect to find these doubly charged ions coexisting with ionized helium.

It is easy to see qualitatively the ionization structures which are possible. Suppose we divide the stellar UV radiation into two photon bands. Band 1 consists of photons of energies in the range 13.6 eV $\leq h\nu \leq$ 24.6 eV. Such photons can ionize only hydrogen and not helium. Band 2 consists of photons of energies $h\nu \geq 24.6$ eV. These photons can ionize both hydrogen and helium. However, as was remarked in Section 5.2.5, the absorption cross-section of hydrogen decreases with increasing photon energy. The absorption cross-section of helium, at a given photon energy, is greater than that of hydrogen at the same photon energy. This difference is enough to compensate for the

TABLE 5.3

Element	First IP	Second IP
H	0.55	–
O	0.55	1.42
N	0.59	1.20
Ne	0.88	1.67

greater abundance of hydrogen relative to helium and, as a result, band 2 photons are mainly absorbed by the helium, provided band 1 photons are also present. If no band 1 photons are present, then band 2 photons are absorbed by both hydrogen and helium. There are then two possibilities for the ionization structure and these are depicted schematically in Figure 5.4(i) and (ii).

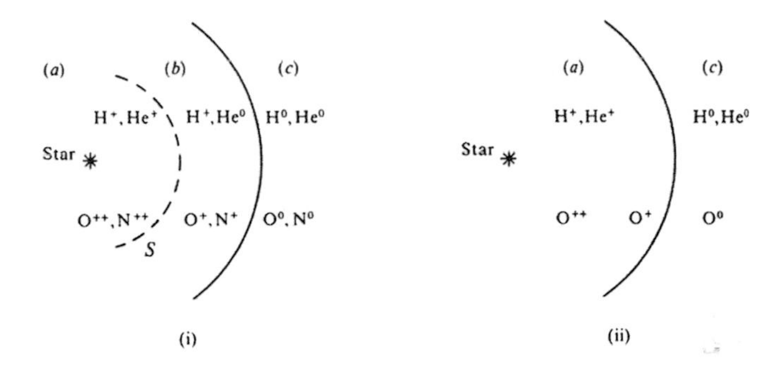

FIGURE 5.4 Ionization stratification in a nebula. (i) Low stellar temperature ($T_* \leq 40000$ K); (ii) High stellar temperature ($T_* \geq 40000$K).

Figure 5.4(i) shows what happens if band 2 photons are absorbed before band 1 photons. S is a surface at which the flux of band 2 photons has become negligible. Interior to S (zone (a)) the gas will contain He^+ and H^+. Exterior to S, helium is neutral. Ionized hydrogen also exists outwards from S to the Strömgren radius and this region (zone (b)) contains neutral helium. Beyond the Strömgren radius, both helium and hydrogen are neutral (zone (c)). From Table 5.3 we can see that zone (a) can contain ions such as O^{++}, N^{++}, and so on. Zone (b) will contain the lower ionization stages, O^+, N^+, and so on. Zone (c) contains only neutral atoms. The ions are clearly spatially stratified. Of course there is a transition zone between zones (a) and (b) which contains a mixture of the higher and lower ionization stages.

Figure 5.4(ii) shows what happens if band 1 photons run out first or (what actually occurs) if band 1 and band 2 photons run out more or less simultaneously. Region (a) contains ionized hydrogen and helium and there is no zone containing neutral helium and ionized hydrogen together. This region contains mainly the higher ionization stages (O^{++} etc.). However, some lower ionization stages are produced near the transition zone between the ionized and neutral gas.

Which situation occurs in practice depends on the temperature of the star. The higher the stellar temperature, the greater is the proportion of stellar UV photons emitted in band 2. As we would expect, the more photons are emitted in this band, the less is the likelihood that photons in band 2 run out before photons in band 1. Thus we anticipate that the situation of Figure 5.4(i) is more likely to occur with cooler exciting stars, whilst the situation of Figure 5.4(ii) is more likely for the hotter stars. This is indeed precisely the case.

Roughly speaking, if $T_* \lesssim 4 \times 10^4$K, then Figure 5.4(i) applies. Figure 5.4(ii) describes the situation for $T_* \gtrsim 4 \times 10^4$K.

5.4 RADIO-FREQUENCY SPECTRA OF NEBULAE

So far, we have confined our discussion to the bright optical lines in nebular spectra, primarily because they dominate the energy balance. However, nebulae radiate in many other ways. For example, a continuum optical spectrum is produced by the recombination of electrons into the various levels. The continuum distribution of photon energies results because the electron energy distribution is continuous.

There are two major contributions to the radio-frequency spectra of nebulae, one giving rise to continuum emission and one to a line spectrum. We now consider them in turn.

5.4.1 The Radio-Frequency Continuum of a Nebula

Bremsstrahlung (or free–free) radiation is produced by the acceleration of electrons in the fields of the positive ions. This radiation is emitted over a wide frequency range but is observed at radio frequencies. In contrast to the optical lines, however, absorption of this radiation within nebulae is important, and the observed continuous spectrum must be interpreted allowing for this effect. Thus one must consider the transfer of the radiation through the nebula.

In order to do this, we define the following quantities: I_ν (the radiation intensity) is the radiant energy in unit frequency interval at frequency ν which crosses unit area per second per unit solid angle; j_ν (the emission coefficient) is the radiant energy emitted by the gas in unit frequency interval at frequency ν per unit volume per second per unit solid angle; κ_ν (the absorption coefficient) in units of m^{-1}.

In plane parallel geometry, the change in I_ν between planes at positions s and $s + ds$ is (cf. Section 2.2.3)

$$dI_\nu = (j_\nu - \kappa_\nu I_\nu)ds$$
$$\text{or} \qquad\qquad (5.37)$$
$$\frac{dI_\nu}{ds} = j_\nu - \kappa_\nu I_\nu.$$

For a system in thermodynamic equilibrium, j_ν and κ_ν are connected by the Kirchhoff relationship (equation (2.19))

$$j_\nu = \kappa_\nu B_\nu(T), \qquad (5.38)$$

where $B_\nu(T)$ is the Planck function at the temperature of the system. The electron gas has a Maxwellian distribution at a temperature T_e. Thus the emission and absorption coefficients for the bremsstrahlung are related by equation (5.38) with T put equal to T_e. Defining the optical depth at frequency ν by

$$d\tau_\nu = \kappa_\nu \, ds, \tag{5.39}$$

equation (5.37) can be written in the form

$$\frac{dI_\nu}{d\tau_\nu} = B_\nu(T_e) - I_\nu. \tag{5.40}$$

The solution of this equation is

$$I_\nu = B_\nu(T_e)(1 - e^{-\tau_\nu}). \tag{5.41}$$

In deriving this solution, we have made the following assumptions: firstly, that T_e is constant throughout the nebula; secondly, that τ_ν is the total optical depth of the nebula, that is, optical depth from one side to the other; thirdly, that $I_\nu = 0$ at $\tau_\nu = 0$, that is, only radiation from the nebula is measured (there is no emission from any other sources, cf. equation (2.21)).

Equation (5.41) has two obvious limiting forms. If $\tau_\nu \ll 1$, then

$$I_\nu \approx B_\nu(T_e)\tau_\nu. \tag{5.42}$$

If $\tau_\nu \gg 1$, then

$$I_\nu \approx B_\nu(T_e). \tag{5.43}$$

We will state, without proof, that for bremsstrahlung, τ_ν has the form

$$\tau_\nu = A\nu^{-2.1} T_e^{-1.35} n_e^2 L, \tag{5.44}$$

where A is a constant and L is the path length through the nebula. The quantity $n_e^2 L$ is an important one and is called the emission measure, ε (its usual units in the astronomical literature are cm^{-6} pc). It is a fixed quantity for a given nebula. At radio frequencies where $h\nu/kT_e \ll 1$, we can use the Rayleigh–Jeans approximation to the Planck function,

$$B_\nu(T_e) \approx 2kT_e\nu^2/c^2 \tag{5.45}$$

where c is the velocity of light. We then find that if $\tau_\nu \ll 1$ (which from equation (5.44) implies high frequencies),

$$I_\nu \propto \nu^{-0.1}. \tag{5.46}$$

At lower frequencies where $\tau_\nu \gg 1$,

$$I_\nu \propto \nu^2. \tag{5.47}$$

The radio-frequency continuous spectrum therefore has the characteristic shape shown in Figure 5.5. Of course the actual numerical values of I_ν as a function of ν depend on the particular parameters (ε and T_e) of the given nebula.

Note one final point with regard to this figure. If we draw straight lines (on a log–log plot) corresponding to equations (5.42) and (5.43) and produce them to meet at some frequency ν_0 (Figure 5.5), then it is evident from these equations that at this frequency $\tau_{\nu_0} = 1$. Thus if T_e is known we can use the derived ν_0 in equation (5.44) (with $\tau_{\nu_0} = 1$) and immediately find the emission measure ε. If, further, we can estimate L then we can estimate n_e, the electron density in the gas. (Strictly, since $\varepsilon \propto n_e^2$, we are deriving a root-mean-square electron density. This qualification is very important for real nebulae, which are extremely inhomogeneous.)

5.4.2 Radio-Frequency Lines from a Nebula

As we have seen (Section 2.2.4), recombination can produce hydrogen atoms in any of the excited levels. Downward radiative transitions from the levels produce a line spectrum. So far we have, for example, mentioned the Balmer lines which are produced by transitions from some level n to the level of principal quantum number n = 2.

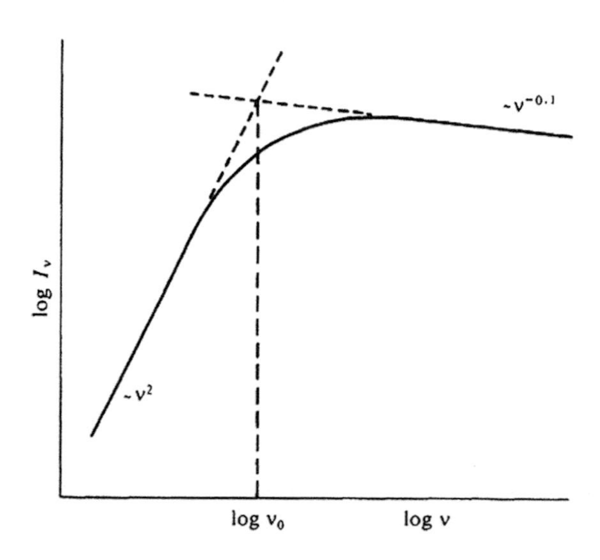

FIGURE 5.5 Intensity against frequency plot for the free–free radiation emitted by a plane parallel nebula of uniform density.

Suppose we consider the general transition between levels of principal quantum numbers n and n'. The frequency of this transition $\nu_{nn'}$ is given by

$$\nu_{nn'} = \mathcal{R}\left(1/n'^2 - 1/n^2\right) \tag{5.48}$$

where $\mathcal{R}$ is the Rydberg constant expressed in frequency units ($\mathcal{R} = 3.3 \times 10^{15}$ Hz). Now, put $n - n' = \Delta n$ and assume Δn is much less than n or n'. We can then write equation (5.48) in the form

$$\nu_{nn'} \approx 2R\Delta n/n'^3. \tag{5.49}$$

Expressing (5.49) in terms of wavelength, we find

$$\lambda_{nn'} \approx 4.5 \times 10^{-8} n'^3/\Delta n \quad \text{m}. \tag{5.50}$$

So if, for example, we put $\Delta n = 1$ and $n' = 100$, then $\lambda_{nn'} = 4.5$ cm, that is, the emitted line lies in the radio region of the spectrum. There are a large number of such transitions which give rise to radio lines. If $n' = n - p$, then it is customary to refer to the line as $(n - 1)\alpha$ line, if $p = 1$; $(n - 2)\beta$ line, if $p = 2$; and so forth.

The radii of these high-n orbits are relatively large. For example, the standard Bohr theory gives the radius of the $n = 100$ orbit as 5.3×10^{-7} m. Since other particles are present in the gas it is of interest to see if such large orbits are well defined. A simple criterion for this to be the case is that the orbit radius should be less than the average inter-particle distance. If, for example, the particle density is 10^8 m^{-3}, the average distance between particles is about 2.1×10^{-3} m. This is much greater than the radii of orbits from which radio wavelength transitions occur and we can indeed consider these orbits to be well defined.

The intensities of these lines can, as for the continuum, be predicted by solving the equation of transfer. However, two additional complications occur. Firstly, these lines are formed at frequencies at which the nebula emits significant amounts of bremsstrahlung. In other words the line sits on a continuum. The transfer equation must take into account the fact that at frequency ν within a line the observed radiation is a mixture of line and continuum. Fortunately, the lines are so narrow that the fractional frequency change across a line is extremely small. Thus, the contribution to the emission from the continuum at frequencies within the line is effectively the same as the continuum emission just outside the line. This can be measured and subtracted from the total emission to give the contribution just from the line.

Perhaps more interestingly, a maser effect can occur in the line. In considering line transfer we must use the absorption coefficient for the radiation. Line photons of energy $h\nu_{nn'}$ can produce two effects. They can either be absorbed by the transition $n' \rightarrow n$ or they can stimulate the transition $n \rightarrow n'$ in an excited atom. The effect of this stimulated emission is to decrease the effective absorption coefficient. This decrease is quite negligible if the line involved lies in the optical region (at gas temperatures typical of diffuse nebulae), but is very important if the transition lies in the radio region. If the population distribution amongst the excited levels follows a Boltzmann distribution (which occurs if thermodynamic equilibrium obtains), the stimulated emission correction is such that the effective absorption coefficient is always positive and absorption decreases the line intensity. But in nebulae, the hydrogen atom populations do not follow a Boltzmann distribution because thermodynamic equilibrium does not occur. (Direct observational evidence for this comes from the relative intensities of optical emission lines, though of course the levels from which they originate are of much lower n than we are considering here. Note that this does not conflict with the assumption of equation (5.38), where we are considering emission by the *free electrons*.) However, it is possible to calculate the distribution of the atoms amongst the excited levels, and it is found that higher levels are slightly overpopulated with respect to lower levels as compared with the case of a Boltzmann distribution. Since the importance of stimulated emission for the transition $n \rightarrow n'$ depends on the upper-level population, stimulated emission is more important than predicted on a Boltzmann distribution. In fact, under certain circumstances, stimulated emission becomes more important than absorption and the effective absorption coefficient becomes negative. Thus we have a maser. These effects are not dramatic – though of some importance in the interpretation of the line intensities. Typically, the masering increases the line intensity by perhaps 20–50%.

5.5 DETERMINING THE PHYSICAL STRUCTURE OF NEBULAE

In the introductory section to this chapter, various observationally determined parameters of nebulae were quoted, in particular, electron temperatures and electron densities. Their values must, of course, be determined from the radiations emitted by the nebulae. We will briefly consider how this can be done.

5.5.1 The Determination of Electron Temperatures

In principle, the simplest method of measuring the electron temperature is to look at the width of an emission line. An emission line is broadened because of the thermal motion of the emitting atoms or ions. Since the typical Doppler shift is $\Delta\lambda \approx v_{\text{th}}/c$, where v_{th} is the thermal velocity ($\equiv (3kT_e/m)^{1/2}$) of the particle, $\Delta\lambda \propto m^{-1/2}$, where m is the particle mass. Obviously the hydrogen lines are the most thermally broadened. However, there is a problem with estimating the temperature simply by measuring $v_{\text{th}}\left(\propto T_e^{1/2}\right)$ from the width of, say, a hydrogen line. As we shall see in Chapter 7, the gas in nebulae may be in a state of motion. The lines can be broadened by the Doppler shifts introduced by mass motions in addition to the thermal motions. Fortunately, we can readily distinguish between the two. Unlike the thermal motions, the broadening due to gas motion is independent of the mass of the emitting particle. In fact, because of the inverse mass dependence, the thermal broadening of, say, an oxygen line is usually quite negligible compared with the broadening due to mass motions. In order to obtain the thermal broadening, we need to measure the width of a hydrogen line and, say, an oxygen line. We use the latter to obtain the broadening due to mass motions and this can be subtracted from the former to give the purely thermal broadening and thus the gas temperature.

An alternative method is to use the relative intensities of forbidden lines from a given ion. Suppose we consider an ion (such as O^{++}) with two levels, say j and k, above the ground level i. The relative intensity of the lines produced by the transition $j \rightarrow i$ and $k \rightarrow i$, say $\mathcal{I}$, is given by (cf. equations (5.32) and (5.33))

$$\mathcal{I} = C_{ij}/C_{ik} = \left(A_{ij}/A_{ik}\right) \exp\left[-\left(\psi_{ij} - \psi_{ik}\right)/kT_e\right]. \tag{5.51}$$

Obviously, we can determine T_e from a measurement of $\mathcal{I}$. Note that we are again using formulae appropriate to the limit of negligible collisional de-excitation from the upper levels. Fortunately, this is a good approximation for the O^{++} lines.

5.5.2 The Determination of Electron Densities

We saw in Section 5.4.1 that one method of estimating n_e comes from a determination of the radio-frequency continuum spectrum of a nebula, but other methods are available. For example, the total energy emitted by a nebula in, say, a Balmer line of hydrogen (e.g. H_β) is proportional to the integral of n_e^2 over the nebular volume, since it depends on the recombination rate. (The constant of proportionality depends on T_e.) Hence a measurement of this energy together with estimates of T_e and the nebular volume enable us to find n_e.

Forbidden lines can also be used to estimate n_e, but we must, in fact, relax our assumption that collisional de-excitation of levels is unimportant. It turns out that at electron densities in the range $10^8\ \text{m}^{-3} \lesssim n_e \lesssim 10^{10}\ \text{m}^{-3}$, the relative intensity of the two lines in the 372.7 nm doublet of O^+ (Figure 5.3(a)) is very sensitive to n_e. It is not very sensitive to T_e because the lines are at nearly the same energy above the ground level. It is possible to calculate the theoretical variation of this ratio as a function of n_e by calculating the level populations as functions of n_e. Observational measurement of the line ratio then gives the electron density.

PROBLEMS

1. An O star $\left(S_* = 10^{49}\text{s}^{-1}\right)$ illuminates a uniform density nebula. How many B stars $\left(S_* = 2 \times 10^{47}\text{s}^{-1}\right)$ would be required to keep the same mass of gas ionized? (Assume that all stellar UV photons are absorbed and that the gas density is the same in both cases.)

2. A hot star excites a uniform cloud of hydrogen of density n_1 m^{-3}. An identical star illuminates a cloud in which the gas is distributed in small clouds of density n_2 m^{-3} ($n_2 > n_1$). Is the mass of ionized gas the same in both nebulae?
3. Show that the average kinetic energy given to a photo-ejected electron is about kT_*, if the star radiates as a black-body at temperature T_*.
4. Two nebulae, excited by identical stars, are observed to have electron temperatures of 9300 K and 7000 K, respectively. The oxygen: hydrogen abundance ratio in the higher temperature nebula is 6×10^{-4}. What is it in the other nebula? (Assume that the gas cools only through emission of lines from the two D levels of O^+.)
5. The observed frequency ν_0 (Figure 5.5) for a certain nebula is 1.96×10^8 Hz. The electron density measured in the nebula is 10^8 m^{-3}. Estimate how many stars each producing 10^{49} UV photons s^{-1} are required to keep the nebula ionized. (Assume that the optical depth at wavelength λ in the free–free continuum is $\tau_\lambda = 4.1 \times 10^{-6} \lambda^{2.1} \varepsilon$, where ε, the emission measure, is expressed in units of cm^{-6} pc and λ is expressed in metres.)
6. An observer measures the profiles of the Hα line and a line from N^+ (at λ = 658.4 nm) in a certain nebula. The measured total half-intensity widths are 0.05 and 0.04 nm, respectively. Find the electron temperature and velocity of random mass motions in the nebula. (Assume that the total half-intensity width of a line of wavelength λ radiated by an ion or atom of mass M is $\delta\lambda = (2\lambda/c)(\ln 2)^{1/2}\left(2kT_e/M + v_t^2\right)^{1/2} v_t$ is the typical velocity of mass motion and c is the velocity of light.)
7. An envelope around a star has a density which varies as $r^{-3/2}$, where r is the radial distance from the star. Show that if the star produces ionizing photons, the ionized part of this envelope is either very close to the stellar surface or effectively at infinity.

6 Introduction to Gas Dynamics

6.1 BASIC EQUATIONS FROM THE CONSERVATION LAWS

6.1.1 Introduction

The mean free paths of the thermal particles in interstellar gas are nearly always short compared with the linear sizes of the regions occupied. A given particle (atom, molecule, or ion) undergoes many collisions before traversing a significant fraction of the region. The particle velocity distributions are therefore Maxwellian, and we can describe them by a gas kinetic temperature, which is usually the same for all species of particle present. The state of the gas can be described in terms of macroscopic properties in addition to temperature, for example, pressure, density, and velocity, which are averages over the properties of many individual particles contained within regions of extent much greater than the mean free paths. Usually, the interstellar gas is in a state of motion and these bulk properties are functions of both position and time. To understand the nature of the motion, we have to derive equations which govern the flow of gases. The exact solution of these equations is usually difficult and will not be discussed. However, a study of them is essential, firstly, to understand the types of phenomena which occur in gas flow, and secondly, to derive the broad principles used to obtain simple models of various events. One such principle which we shall use implicitly on several occasions follows immediately from the basic assumption that the mean free paths are small; this immediately allows us to conclude that the collision of streams of gas does not result in any appreciable diffusion of one gas into another, that is, the two component streams retain their identities after the collision and a distinct boundary exists between them.

6.1.2 Conservation Principles

We derive the equations of gas flow by applying conservation principles to the gas. The quantities that are to be conserved are the mass, the momentum, and the energy. It is usually the equation of energy conservation which demands most care in astrophysical situations. However, we shall discuss here only special cases of this equation.

The conservation principle can be stated in the following general form. Let Π be one of the three properties conserved, and let ν be a fixed volume in space occupied by the gas. Then the rate of increase of Π in the volume ν is equal to the net rate at which Π is advected into the volume ν by the fluid flow plus the net rate at which Π is generated within or over the boundary of volume ν. We shall not use the principle in its most general form since we are mainly interested in deriving simple conclusions from the flow equations. Therefore, we shall assume that the flow has the simplest possible geometry, namely plane parallel, so that the flow variables (density etc.) are functions of a single spatial coordinate x and are constant over planes of given x (see Figure 6.1). The flow is everywhere perpendicular to these planes.

The flow equations can be derived by applying the conservation principle to the small, spatially fixed, volume $\mathrm{d}V$ of unit surface area lying between the planes x and $x + \mathrm{d}x$. Obviously, $\mathrm{d}V \equiv \mathrm{d}x$.

6.1.3 Conservation of Mass

Here the conserved quantity is

$$\Pi \equiv \rho\, dx \tag{6.1}$$

where ρ is the gas density. Assuming that there exist no sources or sinks of material within the volume element, the conservation principle (Section 6.1.2) gives the following equation:

$$\frac{\partial}{\partial t}(\rho\, \mathrm{d}x) = \rho u - (\rho + \mathrm{d}\rho)\,(u + \mathrm{d}u), \tag{6.2}$$

where u is the gas bulk velocity. Since $\mathrm{d}x$ is arbitrary, equation (6.2) can be written (to first order) in the form

$$\frac{\partial \rho}{\partial t} + u\frac{\partial \rho}{\partial x} + \rho\frac{\partial u}{\partial x} = 0. \tag{6.3}$$

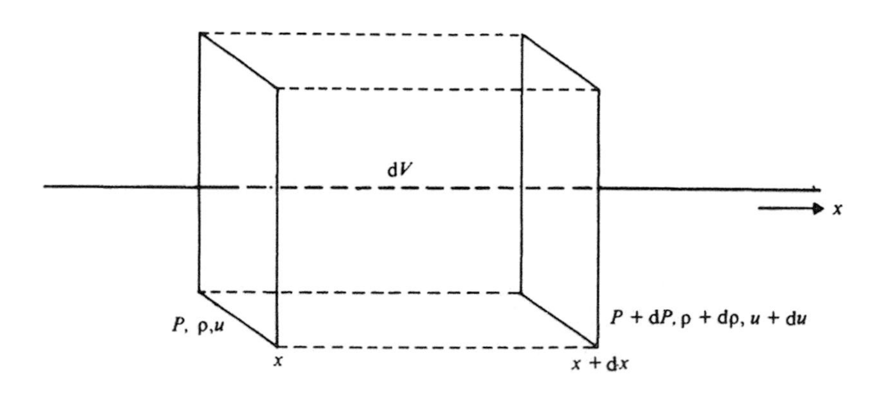

FIGURE 6.1 Control volume for plane parallel flow.

This is the equation for mass conservation (also called the equation of continuity). In equation (6.3) (and in subsequent relationships), the derivatives are partial derivatives. Time derivatives are evaluated at a fixed point in space, whereas spatial derivatives are evaluated at a fixed instant of time. In many astrophysical situations, we have to relax our assumption that there exist no sources or sinks in the flow. For example, matter can be injected into space following explosive events in stars. Alternatively, matter could be lost by accretion onto a black hole. This could be taken into account by adding suitable gain and loss terms to equation (6.3). For time-independent flows where time derivatives vanish (and the spatial partial derivatives become ordinary derivatives), equation (6.3) is integrated to give ρu = constant.

6.1.4 Conservation of Momentum

Here the conserved quantity is

$$\Pi \equiv \rho u \, dx. \tag{6.4}$$

Let us assume that the gas pressure is the only force acting on the gas. In this case a momentum generation term must be included because there are pressure forces acting on the faces of the volume dV. The condition of momentum conservation is that

$$\frac{\partial \rho}{\partial t}(\rho u \, \mathrm{d}x) = \rho u^2 - (\rho + \mathrm{d}\rho)(u + \mathrm{d}u)^2 + P - (P + \mathrm{d}P). \tag{6.5}$$

The first and second terms on the right-hand side of equation (6.5) account for the rate at which momentum enters and leaves the volume, and the third and fourth terms together give the net rate of change of momentum due to the applied force. Using the fact that dx is arbitrary, then, with the help of equation (6.3), equation (6.5) can be written in the form (again, to first order)

$$\frac{\partial u}{\partial t} + u\frac{\partial u}{\partial x} = -\frac{1}{\rho}\frac{\partial P}{\partial x}. \tag{6.6}$$

This equation is often referred to as Euler's equation. In many astrophysical applications, we must include other forces. For example, gravity or magnetic fields may play a role. Often, also, momentum from a radiation field can be transferred to the gas by absorption or scattering. As for the continuity equation (Section 6.1.3), there is a simple integration of the momentum equation (6.6) for time-independent flows. Using the integrated continuity equation, equation (6.6) is integrated to give $P + \rho u^2$ = constant.

An important assumption implicit in the derivation of equation (6.6) is the neglect of viscous forces. We shall now investigate the circumstances when they become important, since, as we shall see, these circumstances may often be met.

In Figure 6.2, Σ is a plane surface drawn perpendicular to the direction of variation of the flow properties. Let the flow velocity vary by an amount U over some characteristic scale length L. This is equivalent to saying that there exists a velocity gradient $\partial u/\partial x \approx U/L$ in the gas. We can take U to be the characteristic gas velocity at Σ. Particles originating at distances less than or of the order of the mean free path, λ, in the gas on either side of Σ can cross Σ before they collide with, and exchange momentum with, other particles. The typical velocity of a particle crossing Σ, relative to the gas flow, is $\bar{c}$, the mean thermal velocity in the gas. From standard kinetic theory, $\bar{c} \approx (kT/m)^{1/2}$, where T is the gas kinetic temperature and m is the particle mass (which for our purposes would be the mass of the hydrogen atom). The net rate at which systematic momentum (i.e. momentum associated with the bulk velocity) is transferred across Σ is given by

$$\rho\bar{c}\left(u+\frac{\partial u}{\partial x}\lambda\right)-\rho\bar{c}\left(u-\frac{\partial u}{\partial x}\lambda\right)=2\rho\bar{c}\frac{\partial u}{\partial x}\lambda \approx 2\rho\bar{c}\frac{U\lambda}{L}. \tag{6.7}$$

In deriving expression (6.7), we have put the systematic velocity change over distance λ equal to $(\partial u/\partial x)\lambda$. Therefore, there is a net change of momentum only when a systematic velocity gradient exists in the gas. Expression (6.7) holds anywhere in the gas since Σ is arbitrary.

Now consider again the volume element $\mathrm{d}V$. The net rate of change of momentum of this element due to this microscopic transfer of momentum must be equal to the difference in rates at which momentum is transferred across the faces at x and $x + \mathrm{d}x$, that is, it is equal to

$$2\rho\bar{c}\lambda\left(\frac{\partial u}{\partial x}+\frac{\partial^2 u}{\partial x^2}\mathrm{d}x\right)-2\rho\bar{c}\lambda\left(\frac{\partial u}{\partial x}\right)\approx 2\rho\bar{c}\lambda\frac{\partial^2 u}{\partial x^2}\mathrm{d}x \approx 2\rho\bar{c}\frac{\lambda U}{L^2}\mathrm{d}x. \tag{6.8}$$

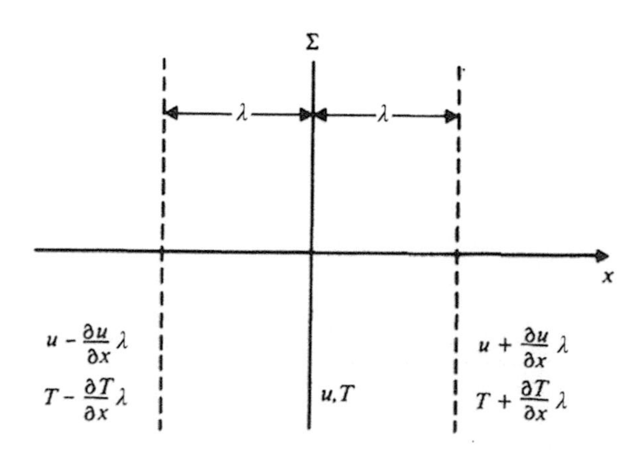

FIGURE 6.2 Arbitrary surface within a region of plane parallel flow.

Here, the change in the velocity gradient between x and $x + dx$ is put equal to $(\partial^2 u/\partial x^2)dx$, and $\partial^2 u/\partial x^2$ is approximated by U/L^2. There is thus a net force, dF_v, acting on the volume dV ($\equiv dx$) as a result of the velocity gradient, given by

$$dF_v \approx 2\rho\bar{c}\frac{\lambda U}{L^2}dV. \tag{6.9}$$

We shall call this the viscous force.

There is, in addition, a net pressure force, dF_p, on the volume element dV, given by

$$dF_p = (P + dP) - P = dP. \tag{6.10}$$

If the pressure changes by the characteristic value P over scale length L then

$$dP \approx (P/L)dx \equiv (P/L)dV. \tag{6.11}$$

Under all circumstances, interstellar gas behaves as a perfect gas. The gas pressure, density, and temperature are related by

$$P = \frac{\rho k T}{\mu m} \tag{6.12}$$

where μ is the mean molecular weight (1 for neutral and $\frac{1}{2}$ for fully ionized hydrogen). To a factor of order unity, $P \approx \rho\bar{c}^2$. Using this expression in equations (6.10) and (6.11), the ratio of the pressure force to the viscous force on the volume element is

$$\frac{dF_p}{dF_v} \approx \left(\frac{\bar{c}}{U}\right)\left(\frac{L}{\lambda}\right). \tag{6.13}$$

As we shall see later (Section 6.2), $\bar{c}$ is about equal to a, the speed at which small perturbations propagate in the gas – the sound speed. The ratio of gas velocity to sound speed (i.e. U/a) will be denoted by M (the Mach number; Section 6.3.3). Hence,

$$\frac{dF_p}{dF_v} \approx \frac{L}{M\lambda}. \tag{6.14}$$

Clearly, $L/\lambda \gtrsim 1$, and usually in interstellar space, $M \lesssim 1$. If $L \gg \lambda$, that is, if the velocity changes over a scale length much greater than the mean free path, $dF_p/dF_v \gg 1$ and viscous forces can be neglected. This can of course be true for $M > 1$ provided L/λ is great enough. Suppose, however, $M \gg 1$ and $L/\lambda \approx 1$, that is, we have a high-speed flow with a very steep velocity gradient. Then $dF_p/dF_v \lesssim 1$ and viscous forces are important. We will examine the consequences of this in Section 6.3. If viscous forces do become important, then they can produce heat. We observe, however, that this process is thermodynamically irreversible. It is therefore associated with an entropy increase. This point will be referred to again in Section 6.3.3.

6.1.5 Conservation of Energy

As mentioned earlier, the energy equation is, in astrophysical situations, often the most complicated of the three conservation equations. However, there exist certain limiting situations for which a general energy equation can be replaced by a simple expression relating various properties of the gas such as pressure and density. We shall discuss two of these.

Adiabatic flow. Suppose that an element of gas neither gains nor loses heat by contact with its surroundings. Thus, we rule out processes such as loss or input of energy by radiation, thermal conduction, and so forth. Any change of the internal energy of the element is affected solely through the element performing work on its surroundings, or vice versa. In this case, the entropy of a gas element remains fixed throughout its flow, and from thermodynamics we know that the pressure and density of the gas are related by the adiabatic relationship

$$P = K\rho^{\gamma}. \tag{6.15}$$

Here γ is the usual ratio of principal specific heats. Since we shall always, in our applications, consider the interstellar gas to be monatomic, $\gamma = 5/3$. K is a constant which is a function of the entropy of the element. If all elements of gas have the same entropy, K is constant everywhere and the flow is called isentropic.

Isothermal flow. In many important situations, the gas can absorb energy from radiation (e.g. from the radiation field of a hot star) and lose energy by radiation (e.g. by forbidden-line emission). An obvious example of this type of situation occurs in the HII regions, which we discussed in Chapter 5. Suppose the gas gains energy at a rate $\overline{\mathrm{G}}$ per unit mass and loses it at a rate $\overline{\mathrm{L}}$ per unit mass. Further, let the timescales associated with the gains and losses be much less than the timescale for any significant flow to occur (i.e. the dynamical timescale). Then the temperature in the gas – as we saw in Section 5.2.8 – is determined by the balance between the heating and cooling rates, that is, by the equation

$$\overline{\mathcal{L}} = \overline{\mathcal{G}} \tag{6.16}$$

Of course, $\overline{\mathcal{G}}$ and $\overline{\mathcal{L}}$ may be functions of such parameters as the density, degree of ionization, and so on. However, the point is that the gas always adjusts to condition (6.16) far more quickly than changes occur due to flow. The simplest possible example again is for the HII regions. In Chapter 5 we showed that, at low densities, equation (6.16) results in a temperature which is both density independent and varies little throughout the whole region of ionized gas. The gas is therefore isothermal at some temperature T and the energy equation has the very simple form

$$T = \text{const.} \tag{6.17}$$

6.2 SOUND WAVES AND THE PROPAGATION OF DISTURBANCES IN GASES

We shall now derive some important results concerning the way in which disturbances are propagated in gases, assuming that the pressure and density are related by

$$P = K\rho^n \tag{6.18}$$

where K is a constant, and n = 5/3 or 1 for adiabatic and isothermal flow, respectively. Equation (6.18) takes the place of an energy-conservation equation. We suppose that the gas is initially at constant pressure P_0 and constant density ρ_0 and that it is at rest, and we allow a small perturbation to be imposed on the gas such that the pressure, density, and velocity take the following respective values:

$$\begin{aligned} P &= P_0 + P_1 \\ \rho &= \rho_0 + \rho_1 \\ u &= u_1. \end{aligned} \tag{6.19}$$

The perturbations P_1, ρ_1, and u_1 will be assumed small, so that when we substitute equations (6.19) into the energy, continuity, and momentum equations we can neglect terms of higher order than linear in the perturbed quantities. Equation (6.18) leads to

$$P_1 = nK\rho_0^{n-1}\rho_1 \equiv n\frac{P_0}{\rho_0}\rho_1. \tag{6.20}$$

We shall define a quantity a_0^2 by

$$a_0^2 = n\frac{P_0}{\rho_0}. \tag{6.21}$$

a_0 has the dimensions of velocity and is a constant. The linearized form of the continuity equation (6.3) is

$$\frac{1}{\rho_0}\frac{\partial \rho_1}{\partial t} + \frac{\partial u_1}{\partial x} = 0. \tag{6.22}$$

Similarly, the linearized momentum equation is

$$\frac{\partial u_1}{\partial t} = -\frac{1}{\rho_0}\frac{\partial P_1}{\partial x}. \tag{6.23}$$

By using equations (6.20) and (6.21), equation (6.23) can be written as

$$\frac{\partial u_1}{\partial t} + \frac{a_0^2}{\rho_0}\frac{\partial \rho_1}{\partial x} = 0. \tag{6.24}$$

Differentiation of equation (6.22) with respect to t, differentiation of equation (6.24) with respect to x, and subtraction of the resulting equations give

$$\frac{\partial^2 \rho_1}{\partial t^2} - a_0^2\frac{\partial^2 \rho_1}{\partial x^2} = 0. \tag{6.25}$$

This is a standard wave equation which has two solutions: a wave of velocity a_0 travelling in the direction of x increasing, and a wave of velocity a_0 travelling in the direction of x decreasing. In general, both waves would be set up by an arbitrary disturbance (although in principle it is possible to choose a disturbance which sets up only one wave). A similar equation can be derived for u_1.

There are two points we should notice about this treatment. Firstly, if – instead of being at rest – the initial velocity of the gas is u_0, then obviously our conclusions still hold provided that we define a new reference frame in which the gas remains stationary. Thus, the disturbance is propagated at velocity a_0 relative to the gas, but in the fixed frame of reference one wave moves with velocity $(u_0 + a_0)$ whilst the other moves with velocity $(u_0 - a_0)$. Secondly, the velocity of the disturbance is expressed in terms of the parameters relating to the undisturbed state of the gas. This is because the disturbances are assumed to be extremely small. If, however, a disturbance is large, then a sound speed can be defined locally anywhere in the disturbance. This sound speed depends on the local values of the pressure and density. From equations (6.18) and (6.21), $a \propto \rho^{(n-1)/2}$, so that a is constant in the isothermal case and proportional to $\rho^{1/3}$ in the adiabatic case. In this latter situation, regions of higher density than average have a higher sound speed than those of lower density. This has extremely important consequences for the gas flow.

We can deduce a very useful general rule of thumb from this discussion. Suppose there exists a distinct region of flow of dimension L, containing a pressure gradient. If there is no means of maintaining this gradient, the flow will evolve so as to remove it. Signals from one region of the flow to another are passed via sound waves having velocity a, and the characteristic time for signals to cross the flow is approximately L/a, which is called the sound-crossing time. This will roughly be the minimum timescale for the pressure gradient to flatten out. Suppose now the general flow in the region, of which L is a part, is subject to some evolutionary changes. If these changes take place slowly compared to the sound-crossing time of region L, then the pressure gradient in region L will tend to be very small. This is an extremely important simplification, and we shall use it in Chapter 7. Although we have obtained it in a rather imprecise way, it is, in fact, supported by solution of the full hydrodynamic equations.

6.3 ADIABATIC SHOCK WAVES AND THEIR PROPERTIES

6.3.1 Shock Waves

An extremely important problem is the behaviour of gases subjected to compression waves. This happens very often in cases of astrophysical interest. For example, a small region of gas suddenly heated by the liberation of energy will expand into its surroundings. The surroundings will be pushed and compressed. We can liken the situation to that in which a piston is pushed into the gas, and use this model to examine the resulting effects.

Figure 6.3(a) shows the initial situation in the model. At time $t = 0$, a piston is pushed into a tube of gas of uniform density ρ_0 and pressure P_0. The piston starts from rest and gradually accelerates into the gas, which we assume behaves adiabatically. The motion of the piston can be thought of as a sequence of small steps of gradually increasing speed. On the first step, a signal travels into the gas at the sound speed and the gas is slightly compressed. On the second step a signal travels into already compressed gas and compresses it further. This signal travels at the higher speed of sound in the gas compressed in stage one. This process can be continued and the pressure (or density) distribution sketched in Figure 6.3(b) will result. However, this process cannot be continued indefinitely: the section of gas nearest to the piston has been both compressed more and accelerated more and has a higher sound speed than its immediate neighbour. This argument can be carried through all the sections right down to the uncompressed gas. The denser back sections try to catch up the front sections and overtake them. This is physically impossible. Obviously, our picture must be modified.

Suppose the catching-up process results in the smoothed-out profile of Figure 6.3(c), which represents schematically the pressure gradient. There will be associated density and velocity gradients. In Section 6.1.4, we showed that when gradients exist in the flow over a distance scale of the order of the mean free path, microscopic processes – particularly those giving rise to viscous forces – must be included in the momentum equation. (Thermal conduction also becomes important in the energy equation under these circumstances.) Our conclusions regarding the behaviour of the compression process were derived from the solution of the flow equations, which specifically excluded these processes. Obviously, arguments dependent on these equations can only be carried up to the point at which the pressure gradient in Figure 6.3(c) exists over a length scale of the order of the mean free path. Then microscopic processes act to prevent the pressure profile becoming steeper (or even overturning) and, since the mean free path is so short, the discontinuous profile given in Figure 6.3(d) can be thought of as the eventual result. This surface at which the pressure (and also density and velocity) changes discontinuously is called a shock wave.

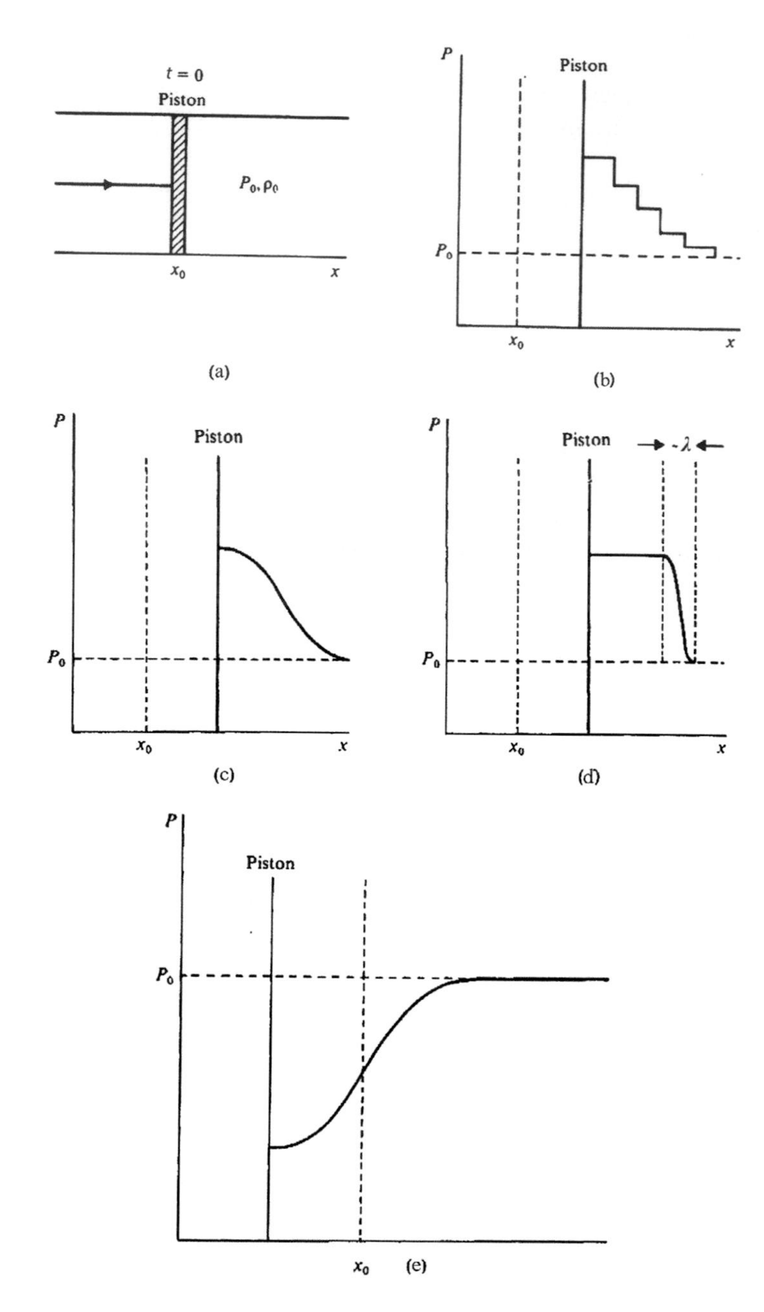

FIGURE 6.3 (a)–(d) To illustrate the steepening of a pressure pulse to form a shock wave; (e) the pressure profile produced by pulling a piston out of a tube.

One final point to notice concerns the effects produced by pulling the piston to the left in Figure 6.3(a). The gas nearest to the piston now has the lowest density and also is moving most quickly (to the left). There is no tendency for the high-density regions to catch up the low-density regions and a smooth flow results with the pressure (or density) gradient shown schematically in Figure 6.3(e). This is called a *rarefaction wave.* The head of the wave (immediately adjacent to the undisturbed gas) moves into the undisturbed gas at the local sound speed.

6.3.2 Properties of Shock Waves: The Jump Conditions

We now look at the properties of shock waves, and, in particular, investigate conservation of quantities across the shock – the so-called jump conditions. Fortunately, the thickness of a shock is so small (comparable to the mean free path, λ) that we can regard it as a discontinuity. Because of this we can always assume it to be locally plane. Further, because the shock is thin the gas flows through it so quickly that changes in the flow conditions cannot affect the details of the transition. The flow into and out of the shock can therefore be regarded as time independent, and the relationships between conditions on each side are also time independent. We can therefore take a frame of reference in which the shock is stationary and will do so for the moment in order to establish the jump conditions. Later on (Section 6.3.5) we shall discuss the simple transformations used to write these conditions in terms of velocities in a reference frame in which the shock moves. Figure 6.4 shows schematically the shock transition. Gas at pressure P_0 and density ρ_0 enters the shock with velocity u_0 relative to the shock. This is the upstream gas. Similarly, downstream gas leaves the shock with pressure P_1, density ρ_1, and velocity u_1 relative to the shock. Conservation principles relate conditions on each side of the shock.

Mass conservation states that the mass inflow rate per unit area of shock front is equal to the mass outflow rate per unit area. Therefore,

$$\rho_0 u_0 = \rho_1 u_1 = \phi \qquad \text{(constant).} \tag{6.26}$$

Momentum conservation follows by application of Newton's Second Law. In unit time, mass $\rho_0 u_0$ enters unit area of front with momentum $(\rho_0 u_0)u_0$. This mass emerges with momentum $(\rho_0 u_0)u_1$. The change of momentum is equal to the impulse given by the net force. Thus,

$$\rho_0 u_0 u_1 - \rho_0 u_0^2 = P_0 - P_1. \tag{6.27}$$

On using equation (6.26), equation (6.27) can be written as

$$P_0 + \rho_0 u_0^2 = P_1 + \rho_1 u_1^2 = \zeta(\text{constant}). \tag{6.28}$$

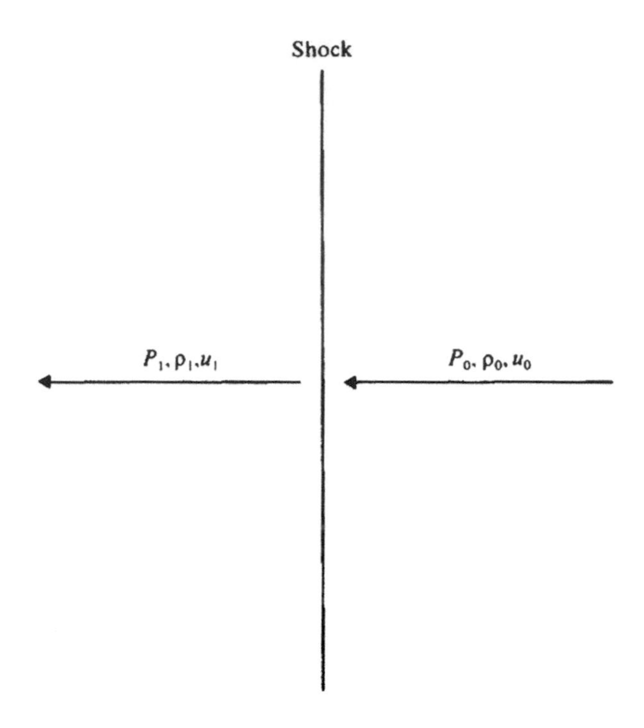

FIGURE 6.4 The flow variables are defined on each side of a shock. Velocities are given in the rest frame of the shock.

We have already seen (Section 6.1.4) that energy can be transformed from one form to another by microscopic processes (e.g. viscosity) within the shock front itself. However, there are no net energy losses or gains from a given mass of gas crossing the discontinuity. We can now write down energy conservation in the following way. In unit time, the energy, E_0, of the gas entering unit area of the shock is

$$E_0 = \rho_0 u_0 \left(\frac{1}{2} u_0^2 + \frac{3}{2} \frac{P_0}{\rho_0} \right). \tag{6.29}$$

The first term is the kinetic energy per unit mass and the second term is the internal energy per unit mass. (We are assuming a ratio of principal specific heats γ of 5/3.) The rate E_1 at which energy leaves unit area of the shock is, similarly,

$$E_1 = \rho_1 u_1 \left(\frac{1}{2} u_1^2 + \frac{3}{2} \frac{P_1}{\rho_1} \right). \tag{6.30}$$

There is a difference between E_1 and E_0 equal to the net rate of working on the gas by the pressure difference. Thus,

$$E_1 - E_0 = P_0 u_0 - P_1 u_1. \tag{6.31}$$

Using equations (6.29)–(6.31), together with (6.26), gives the energy balance condition

$$\frac{1}{2}u_0^2 + \frac{5}{2}\frac{P_0}{\rho_0} = \frac{1}{2}u_1^2 + \frac{5}{2}\frac{P_1}{\rho_1} = \xi \quad \text{(constant)}. \tag{6.32}$$

The quantity $\frac{1}{2}u^2 + \frac{5}{2}P/\rho$ is the specific total energy of the gas. (It is sometimes called the stagnation enthalpy.) Equations (6.26), (6.28), and (6.32), relating the upstream and downstream values of the flow variables, are called the Rankine–Hugoniot conditions, and we note that equation (6.26) is the integrated time-independent continuity equation, and equation (6.28) likewise is the integrated time-independent momentum equation.

6.3.3 What the Rankine–Hugoniot Conditions Tell Us

We have found that the following quantities are conserved across a shock wave:

$$\phi \equiv \rho u \quad \text{(mass flux)} \tag{6.33}$$

$$\zeta \equiv P + \rho u^2 \quad \text{(momentum flux)} \tag{6.34}$$

$$\xi \equiv \frac{1}{2}u^2 + \frac{5}{2}\frac{P}{\rho} \quad \text{(specific total energy)}. \tag{6.35}$$

The actual flow variables, P, ρ, and u, change discontinuously but in such a way as to conserve the above quantities. A convenient way of investigating the changes across a shock wave is to consider the properties of a flow obeying equations (6.33)–(6.35). Further, assume that the gas everywhere obeys an adiabatic equation of state with a local sound speed a defined by

$$a^2 = \frac{5}{3}\frac{P}{\rho}. \tag{6.36}$$

Let us introduce a reference velocity $\bar{u}$, which is defined by

$$\bar{u} = \zeta/\phi. \tag{6.37}$$

We can then manipulate the momentum equation (6.34) to give us the quadratic equation

$$u^2 - u\bar{u} + \frac{3}{5}a^2 = 0. \tag{6.38}$$

This equation has two roots. Since it is derived from the conservation equations, the two roots represent the values of velocity on either side of the shock front. For the moment, however, we use equation (6.38) to express a in terms of u and $\bar{u}$. If we do this, the specific total energy can be written as

$$\xi = \frac{1}{2}u^2 + \frac{3}{2}a^2 = u\left(\frac{5}{2}\bar{u} - 2u\right). \tag{6.39}$$

The specific internal energy is just

$$e_I = \frac{3}{2}\frac{P}{\rho} = \frac{3}{2}u(\bar{u} - u). \tag{6.40}$$

We shall also need the Mach number M, which we defined as the ratio of the gas velocity to the adiabatic sound velocity. Thus,

$$M^2 = \frac{u^2}{(5/3)(P/\rho)} = \frac{3}{5}\frac{u}{(\bar{u} - u)}. \tag{6.41}$$

(This definition of M is quite general. However, its value obviously depends on the reference frame in which the gas velocity is measured.) It is convenient to write the velocity in terms of the reference velocity $\bar{u}$ by introducing the variable $\eta \equiv u/\bar{u}$. We can then write

$$\varepsilon_T \equiv \xi/\bar{u}^2 = \eta\left(\frac{5}{2} - 2\eta\right) \tag{6.42}$$

$$\varepsilon_T \equiv e_I/\bar{u}^2 = \frac{3}{2}\eta(1 - \eta). \tag{6.43}$$

The specific kinetic energy in dimensionless form is obviously

$$\varepsilon_K = \frac{1}{2}\eta^2. \tag{6.44}$$

Similarly,

$$M^2 = \frac{3}{5}\frac{\eta}{(1 - \eta)}. \tag{6.45}$$

Note that η must have a value less than one.

These various quantities ε_T, ε_I, ε_K, and M have been plotted (in arbitrary units) as functions of η in Figure 6.5(a), (b), (c), and (d) respectively. Now, we know that the specific total energy, ε_T, is constant across a shock front. Thus, the shock transition is represented in Figure 6.5(a) by the jump from conditions at point ‘a’ to conditions at point ‘b’ (or vice versa). Obviously the line joining ‘a’ and ‘b’ is horizontal and thus the value of η_b allowable for a given η_a is fixed. So we can put in the corresponding jumps in Figures 6.5(b)–(d) for our given

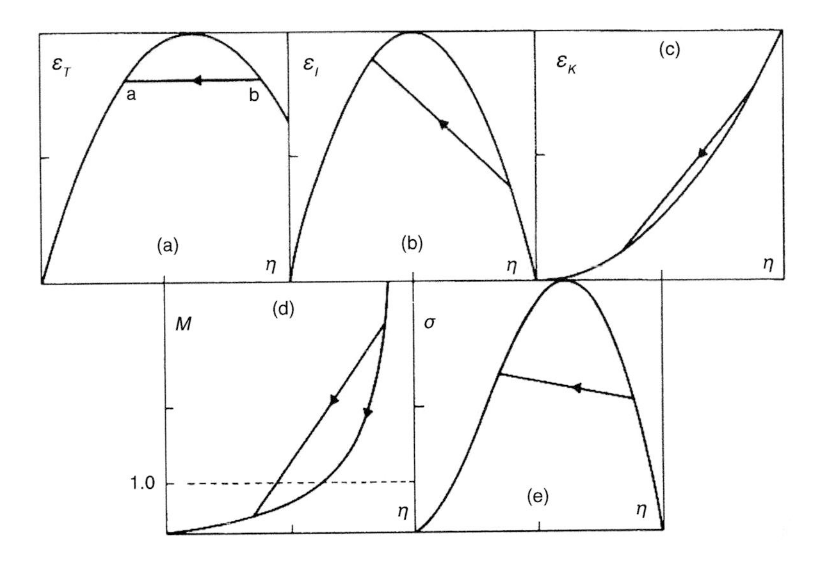

FIGURE 6.5 The variation of dimensionless parameters in steady one-dimensional flow. (a) Specific total energy, (b) specific internal energy, (c) specific kinetic energy, (d) Mach number, and (e) entropy function.

η_a and η_b. Note that the lines connecting 'a' and 'b' are no longer horizontal in these figures and therefore ε_I, ε_K, and M change across a shock transition.

We now must find some condition which will determine whether a shock transition can be in either direction or only in one direction. We appeal to the principle of entropy increase. Since dissipative processes (e.g. viscosity) occur, the entropy of a particle can only increase across the shock. From thermodynamics, the specific entropy, s, of an ideal monatomic gas is given by

$$s = A \ln\left(P\rho^{-5/3}\right) + B \tag{6.46}$$

where A and B are constants. Write this in the form

$$s' \equiv (1/B')\exp(s/A) = P\rho^{-5/3} \tag{6.47}$$

where $\ln B' = B/A$. Obviously, s' is a monotonic function of s and we can consider variations in s' instead of variations in s. From equations (6.33) and (6.40), and using the definition of η,

$$\sigma \equiv s'\phi^{2/3}/\bar{u}^{8/3} = \eta^{5/3}(1-\eta) \tag{6.48}$$

so that σ is a dimensionless quantity, which is a monotonic function of specific entropy whose behaviour can be taken to be that of the specific entropy. Figure 6.5(e) shows σ (in arbitrary units) as a function of η. The transition line 'ab' for the same jump as shown in Figures 6.5(a)–(d) is drawn. Immediately we see that the specific entropy at 'a' is greater than that at 'b'. (This statement is quite general, as an examination of all possible jumps would show.) Thus the shock transition can take place only in the direction 'b → a'.

We can now draw several qualitative conclusions from Figures 6.5(b)–(d). Figure 6.5(b) shows that the specific internal energy of the gas increases across the shock. Figure 6.5(c) shows that this is obtained at the expense of the specific kinetic energy. Figure 6.5(d) clearly indicates that the Mach number of the upstream flow is greater than unity, whilst the downstream Mach number is less than unity. Thus, gas enters a shock supersonically and leaves it subsonically.

6.3.4 Density, Pressure, and Temperature Changes across a Shock

We can now easily derive the relationships which connect quantities on either side of the shock wave. To do this, let us write equation (6.39) in the following form:

$$u^2 - \frac{5}{4}u\bar{u} + \xi/2 = 0. \tag{6.49}$$

Then, for given values of ξ and $\bar{u}$ (i.e. ϕ and ζ), the two roots of this equation (say, u_0 and u_1) give the upstream and downstream gas velocities, respectively. Since (6.49) is a quadratic equation, the sum of the two roots is

$$u_0 + u_1 = \frac{5}{4}\bar{u}. \tag{6.50}$$

Let M_0 ($\equiv u_0/a_0$) and M_1 ($\equiv u_1/a_1$) be the upstream and downstream Mach numbers, respectively. Then from equation (6.41)

$$M_0^2 = \frac{3}{5}\frac{u_0}{(\bar{u} - u_0)}. \tag{6.51}$$

Elimination of $\bar{u}$ from equations (6.50) and (6.51) gives

$$\frac{u_1}{u_0} = \frac{M_0^2 + 3}{4M_0^2}. \tag{6.52}$$

Instead of deriving other relationships in their most general form, we will consider the special case of strong shock waves which are defined by $M_0 \gg 1$. Then equation (6.52) immediately simplifies to give

$$\frac{u_1}{u_0} = \frac{1}{4}. \tag{6.53}$$

From the continuity equation (6.33), we find immediately that the density ratio across the shock is

$$\frac{\rho_1}{\rho_0} = \frac{4}{1}. \tag{6.54}$$

The compression is limited to this value even if M_0 is made arbitrarily large. The physical reason for this is that the change in kinetic energy goes entirely into translational energy of the particles of shocked gas, that is, increases the pressure. This opposes compression. As one might expect the compression ratio is higher for gases (e.g. molecular gases) which have non-translational degrees of freedom.

Since $M_0 \gg 1$, the thermal pressure of the gas ahead of the shock is negligible, and the pressure change across the shock is extremely large. The pressure in the shocked gas is given by (equation (6.28) with P_0 neglected)

$$P_1 = \rho_0 u_0^2 - \rho_1 u_1^2. \tag{6.55}$$

Using equations (6.53) and (6.54),

$$P_1 = \frac{3}{4}\rho_0 u_0^2. \tag{6.56}$$

Since the gas is a perfect gas, it obeys the equation of state

$$P = \frac{\rho k T}{\mu m}. \tag{6.57}$$

Hence, from equations (6.54), (6.56), and (6.57), the temperature behind the shock is

$$T_1 = \frac{3}{16}\frac{\mu m}{k} u_0^2. \tag{6.58}$$

6.3.5 Results in a Fixed Frame of Reference

We shall usually require the results of Section 6.3.4 expressed in forms in which velocities are measured in a fixed frame of reference, that is, one in which the shock is moving. For example, when considering motions produced by the interaction of stars and interstellar gas, it is usually most convenient to express velocities in a frame of reference in which the star is stationary.

To make the transformation, let V_s be the shock velocity in the fixed frame. Let v_0 and v_1 respectively be the upstream and downstream gas velocities, again measured in the fixed frame. The necessary transformation is effected by the relationships

$$u_0 = v_0 - V_s \tag{6.59}$$

and

$$u_1 = v_1 - V_s. \tag{6.60}$$

In our application of these results in Chapter 7, we shall deal only with cases where $V_s \gg v_0$. Obviously, equation (6.54) remains unaffected by the change of reference frames. Equation (6.53) becomes

$$\frac{V_s - v_1}{V_s} = \frac{1}{4} \tag{6.61}$$

giving

$$v_1 = \frac{3}{4} V_s. \tag{6.62}$$

Thus, the gas behind the shock moves in the same direction as the shock but with three-quarters of the shock speed. Equations (6.56) and (6.58) take the respective forms

$$P_1 = \frac{3}{4} \rho_0 V_s^2 \tag{6.63}$$

and

$$T_1 = \frac{3\mu m}{16k} V_s^2. \tag{6.64}$$

Finally, the specific internal and kinetic energies of the shocked gas are given respectively by

$$e_{I_1} = \frac{3}{2} \frac{P_1}{\rho_1} = \frac{9}{32} \rho_0 V_s^2 \tag{6.65}$$

and

$$e_{K_1} = \frac{1}{2} v_1^2 = \frac{9}{32} \rho_0 V_s^2. \tag{6.66}$$

These results will be needed in Chapter 7.

6.4 RADIATING SHOCK WAVES

6.4.1 Cooling Processes in Shock-Excited Gas

When gas goes through a shock wave, it is heated. In general, heated gases radiate and the radiation removes energy from the gas. The temperature behind a shock wave depends on the shock velocity and the molecular weight of the gas (equation (6.64)), and this temperature largely determines the physical mechanisms by which the shocked gas radiates.

We will often be interested in circumstances where both the upstream and downstream gas is ionized or where the shock speed is sufficiently high that ionization can result from high-enough post-shock temperatures. The mean thermal energy of a particle ($\approx 3kT/2$) is equal to the ionization potential of hydrogen (=13.6 eV) at a temperature $T \approx 1.6 \times 10^5$ K. Using equation (6.64) (with $\mu = \frac{1}{2}$) shows that this temperature is reached behind a shock wave with a velocity $V_s \approx 100$ km s^{-1}; we would then expect hydrogen to be fully ionized behind shocks with velocity $V_s \gtrsim 100$ km s^{-1}. In fact, hydrogen is largely ionized at gas temperatures in excess of about 4×10^4 K, that is, for shock velocities $V_s \gtrsim 50$ km s^{-1}. This is because the high-energy Maxwellian tail in the electron energy distribution at the lower temperature can more or less fully ionize the hydrogen. Obviously, the extent to which other elements are ionized depends also on the shock velocity. For example, oxygen is ionized to O^+ behind shocks which ionize hydrogen, but shock velocities of about a factor of 2 higher ($V_s \gtrsim 100$ km s^{-1}) are required to ionize oxygen largely to O^{++}.

The post-shock gas is composed of hot electrons, protons, and various ions. Such gas radiates well by a variety of processes. We emphasize again here (cf. Chapter 5) that, because the collision times between particles in the gas are very short compared to the timescales to lose significant thermal energy by radiation, taking energy out of any one component of the gas is equivalent to cooling the gas as a whole.

We have already met two relevant cooling processes in Chapter 5: bremsstrahlung cooling and forbidden-line cooling. We must, however, consider additional processes because the gas temperature behind shock waves depends on the shock velocity and is not limited to the narrow temperature range of photoionized gas. We will now briefly consider these additional processes.

As long as some neutral hydrogen remains in the gas, cooling can occur by two inelastic collision processes involving electrons; these are the excitation of a hydrogen atom from the $n = 1$ level to a higher level followed by radiative decay and the extension of this to collisional ionization for electrons of sufficient energy. These processes in fact dominate cooling at electron temperatures of about 10^5 K.

Forbidden-line cooling (Chapter 5) is dominant at temperatures between 10^4 and 10^5 K, but once the temperature exceeds this upper limit, the ions (O^+, O^{++}, etc.) which have metastable levels in their ground-state configurations become very rare. The gas now starts to contain more highly ionized species such as C^{3+} and O^{4+}. Again, these ions have energy levels which can be collisionally excited and which subsequently decay radiatively. But there is a big difference: these levels are not metastable and the radiative transitions have very high transition probabilities. The lines produced are called 'resonance' lines and appear in the far UV and soft X-ray regions of the spectrum. Cooling by these resonance lines is dominant in the approximate temperature range 10^5 K $\lesssim 10^8$ K. Once the temperature exceeds this upper limit, ions producing the resonance lines no longer exist; essentially the gas consists of electrons, protons, and bare nuclei of, for example, O and C. Then the only cooling process which is significant is the emission of bremsstrahlung.

In order to examine the consequences of cooling behind shock waves, we write the cooling rate per unit volume in shocked gas as

$$\mathcal{L} = n^2 \Lambda(T_e)\,\mathrm{W\ m^{-3}}. \tag{6.67}$$

The temperature function, $\Lambda\ (T_e)$, is sketched (in arbitrary units) in Figure 6.6. We summarize the main cooling processes contributing to $\Lambda(T_e)$ in Table 6.1 (the letters defining the regions are shown in Figure 6.6).

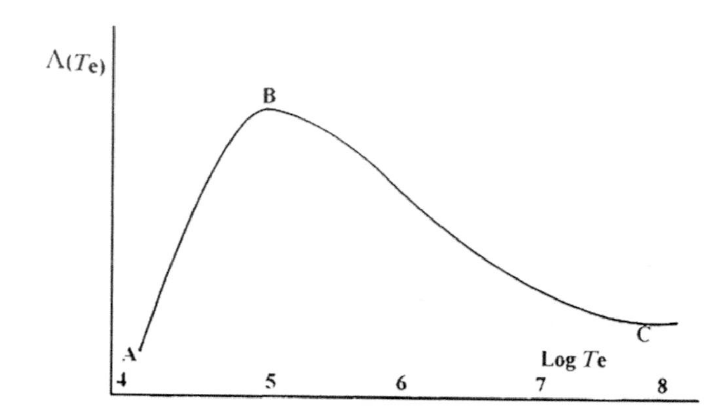

FIGURE 6.6 The cooling function $\Lambda(T_e)$ for shocked interstellar gas.

TABLE 6.1

Cooling processes in various temperature ranges

Region	Approximate temperature range	Main cooling process	Spectral region of the radiation
A → B	$5 \times 10^3\ \mathrm{K} \lesssim T \lesssim 10^5\ \mathrm{K}$	Forbidden lines	Optical; IR
B	$T \approx 10^5\ \mathrm{K}$	H excitation/ionization	Optical; UV
B → C	$10^5\ \mathrm{K} \lesssim T \lesssim 10^8\ \mathrm{K}$	Resonance lines	Far UV; soft X-ray
C onwards	$T \gtrsim 10^8\ \mathrm{K}$	Bremsstrahlung	Radio

Two additional comments need to be made. The cooling function $\Lambda(T_e)$ can be calculated in a consistent fashion by assuming that the abundances of the ions contributing to the cooling are determined by the equilibrium between the rates at which ions are produced and destroyed (cf. equation (5.10)). In this case, however, ions are produced either by the collisional ionization of a more lowly ionized ion or the recombination of a more highly ionized ion. Photoionization plays no role. Secondly, the cooling rate depends upon n^2 since it depends upon the number densities of both the ions and the impacting particles.

The section of the cooling curve B → C (Figure 6.6) is of considerable interest for two reasons. Shocks with velocities ($V_s \approx$ 80–2700 km s^{-1}) which heat gas to such temperatures are rather common (cf. Chapter 7). Also, the cooling rate decreases as the temperature increases. This is simply because there are fewer ions around for cooling as the temperature increases. The major consequence of this behaviour is that gas which starts to cool somewhere between B and C is thermally unstable; that is, the more it cools, the faster it cools. This has considerable significance for the late stages of evolution of supernova remnants (Section 7.3.4).

6.4.2 Cooling Times and Cooling Distances behind Shock Waves

We now look at the situation (Figure 6.7) where gas passes through a strong adiabatic shock S and where radiative cooling occurs after the gas has been shocked. The flow variables across the adiabatic shock are of course related by the standard Rankine–Hugoniot conditions (Section 6.3.3). We define the cooling time, t_c, in the shocked gas as the time taken for unit mass of gas to radiate away most of the thermal energy it gained on passage through the shock. Similarly, we define the cooling length l_c, as the distance travelled by this mass of gas relative to the shock, during this cooling time. We will ignore the density and temperature changes which must occur as this gas cools, since for our purposes, the inclusion of these changes – which must affect **L** (equation 6.67) – is unimportant.

We write the cooling rate per unit mass in the gas immediately behind the shock as

$$\mathcal{L}' \equiv \frac{\mathcal{L}}{n_s \bar{m}} = \frac{n_s \Lambda(T_s)}{\bar{m}} \text{ W kg}^{-1} \tag{6.68}$$

where $\bar{m}$ is the mean mass per particle and subscript s indicates the value of the parameter immediately behind shock S. The specific thermal energy behind the shock e_{I_s} is given by equation (6.65) as $3kT_s/\bar{m}$. The cooling time is then defined as

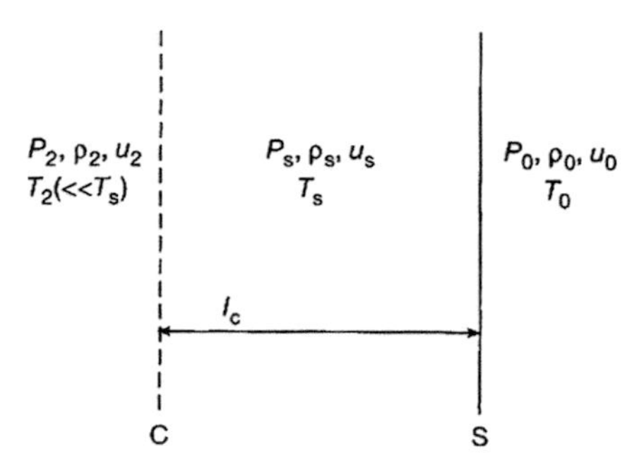

FIGURE 6.7 Schematic representation of a radiating shock wave.

$$t_c \equiv \bar{m}e_{I_s}/n_s\Lambda(T_s) = 3kT_s/n_s\Lambda(T_s). \tag{6.69}$$

In the shock frame, the post-shock gas has velocity $V_s/4$ (see equation (6.53). The cooling length is then

$$l_c \equiv V_s t_c/4 = 3kT_s V_s/4n_s\Lambda(T_s). \tag{6.70}$$

If n_0 is the density ahead of the strong shock, $n_s = 4n_0$ (see equation (6.54).

In order to go further, we would have to put in the calculated form of Λ (T_s), but in view of the significance of the region B → C, we will concentrate on that region only. An excellent functional representation of $\Lambda(T)$ in this region is $\Lambda(T) \propto T^{-1/2}$. Hence, using equation (6.69), $t_c \propto T_s^{3/2}$. Since $T_s \propto V_s^2$ (equation (6.64), $t_c \propto V_s^3$. Further, $l_c \propto V_s t_c$ and so $l_c \propto V_s^4$. Thus, cooling times and cooling lengths behind strong shocks under rather usual conditions are very sensitive to the shock velocity.

6.4.3 Isothermal Shock Waves

An important application of the study of radiating shock waves is to the case where the shocked gas returns eventually to the temperature T_0, which it had before it entered the shock wave (i.e. $T_0 \approx T_2$, Figure 6.7). After passing through the radiating region, this temperature is reached at some surface C (Figure 6.7). If the shock S and surface C are close enough, the gas flows so quickly between S and C that this region can be regarded as thin and the flow as time independent. This is equivalent to saying that t_c is small compared to the dynamical time of the system and that l_c is consequently small compared to the scale size of the system. Surfaces S and C can then be considered to form one surface across which the density and other parameters change discontinuously, but across which the *temperature* does not change. This is sometimes called an isothermal discontinuity or an isothermal shock wave. Note though that it is really an adiabatic shock wave plus a cooling region.

Let subscripts 2 refer to the conditions at surface C and beyond, that is, where $T = T_0$; subscripts 0 refer to the upstream gas (Figure 6.7). The continuity and momentum conservation equations are as for a standard shock; the change occurs in the energy equation. Instead of summing over the various forms of energy, which would now include radiation, we simply write

$$T_0 = T_2 = \text{constant}. \tag{6.71}$$

The sound speed is now (equation (6.21)

$$c^2 = \frac{P}{\rho} = \frac{kT}{\mu m} \tag{6.72}$$

where we use c, the isothermal sound speed to distinguish it from a, the adiabatic sound speed. If c_0 is the sound speed at temperature T_0, the condition of isothermality across the discontinuity can be written as

$$\frac{P_0}{\rho_0} = \frac{P_2}{\rho_2} = c_0^2. \tag{6.73}$$

Again, we shall assume that the shock is strong, that is, $\rho_0 u_0^2 \gg P_0$. Using equation (6.73), the momentum equation (6.28) then becomes

$$\rho_2 c_0^2 = \rho_0 u_0^2 - \rho_2 u_2^2. \tag{6.74}$$

Hence from the continuity condition,

$$u_2^2 - u_2 u_0 + c_0^2 = 0. \tag{6.75}$$

The solution of this is

$$u_2 = (u_0/2)\left(1 \pm \sqrt{1 - 4c_0^2/u_0^2}\right). \tag{6.76}$$

Since we have assumed the shock to be strong, $u_0 \gg c_0$ and thus

$$u_2 = (u_0/2)\left[1 \pm \left(1 - 2c_0^2/u_0^2\right)\right]. \tag{6.77}$$

The positive sign gives the trivial solution $u_2 \approx u_0$. Since we know compression must occur across the adiabatic shock, we must take the negative sign and obtain

$$u_2 \approx c_0^2/u_0. \tag{6.78}$$

The upstream Mach number, which is now defined with respect to the isothermal sound speed is

$$M_0 = u_0/c_0. \tag{6.79}$$

Thus we can write, using the continuity equation and equation (6.78),

$$u_0/u_2 = \rho_2/\rho_0 = M_0^2. \tag{6.80}$$

We now see that the compression across a strong isothermal shock depends on the upstream Mach number, in contrast to the case of a strong adiabatic shock for which the compression is limited to a factor 4. The physical reason for this is that, in order that the gas remains isothermal, internal energy has to be radiated away. This internal energy would otherwise have limited the compression.

The post-shock pressure is now given by

$$P_2 = \rho_2 c_0^2 = \rho_0 u_0^2. \tag{6.81}$$

As before, we will need these results expressed in terms of velocities measured in a fixed frame of reference. The transformation equations (6.59) and (6.60) (with v_1 replaced by v_2) are used and we will further assume $V_s \gg v_0$. The compression ratio ρ_2/ρ_0 remains unchanged. The velocity ratio is

$$\frac{V_s}{V_s - v_2} = M_0^2. \tag{6.82}$$

Thus

$$v_2 = V_s(1 - 1/M_0^2) \approx V_s. \tag{6.83}$$

From equation (6.81), the post-shock pressure is

$$P_2 = \rho_0 V_s^2. \tag{6.84}$$

Note now that equation (6.83) shows that the gas behind a strong isothermal shock moves in the same direction and with the same speed as the shock.

6.5 SHOCK WAVES WITH MAGNETIC FIELDS

6.5.1 Strong Coupling

We have earlier (Section 2.5) noted that a magnetic field permeates the interstellar medium. This magnetic field may be extremely well coupled to the interstellar gas – so well, in fact, that it is then usually referred to as being *frozen* into the gas. Consequently, when interstellar gas is compressed, for example by a shock wave, the magnetic field is also compressed. The properties of shock waves are modified as a result. In fact, if magnetic fields are present, it is the electron gyration radius that determines the effective shock thickness. However, this is generally small enough so that the discussion in Section 6.3 remains valid.

We will here make only a few simple remarks. We assume that the magnetic field is parallel to the shock (Figure 6.8). Upstream of the shock, the magnetic field has strength B_0 and downstream B_1.

The strength of a magnetic field is proportional to the number of field lines crossing unit area of a surface perpendicular to the field direction. The appropriate element of area is therefore parallel to the x direction (Figure 6.8). We consider a unit area face fluid element which has a pre-shock volume $dV_0 \equiv dx_0$ and post-shock volume $dV_1 \equiv dx_1$. Because the field is frozen into the gas, the number of field lines through the fluid element is conserved. They cut area dx_0 before crossing the shock and area dx_1 after crossing the shock. Hence the field conservation condition is

$$B_0 dx_0 = B_1 dx_1. \tag{6.85}$$

Conservation of mass for the fluid element is

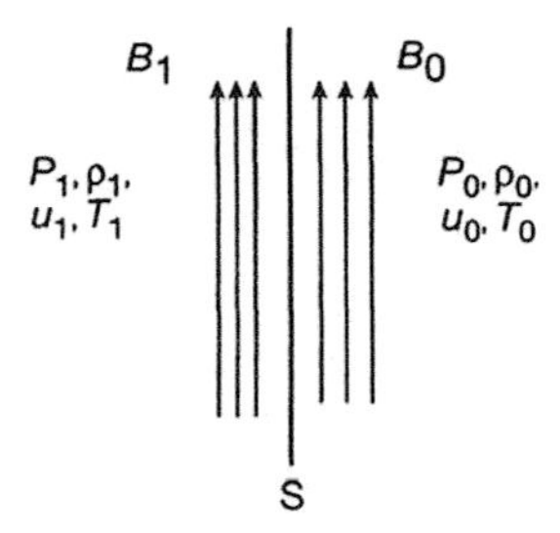

FIGURE 6.8 Schematic representation of a magnetic shock. The field lines are perpendicular to the x direction.

$$\rho_0 \mathrm{d}V_0 \equiv \rho_0 \mathrm{d}x_0 = \rho_1 \mathrm{d}x_1 \equiv \rho_1 \mathrm{d}V_1. \tag{6.86}$$

Therefore,

$$\frac{B_0}{\rho_0} = \frac{B_1}{\rho_1}. \tag{6.87}$$

As we expect, therefore, the field strength increases across the shock. If we were to carry out a full analysis of magnetized shock waves, we would have to write down appropriate forms of the Rankine–Hugoniot conditions. For example, in the momentum equation we would have to account for the pressure of the magnetic field (which acts perpendicular to the field lines). In the energy equation, we would need to include the magnetic energy density. We will confine our remarks to the case of strong shocks where the upstream magnetic field pressure and energy density are small compared to the upstream gas pressure and kinetic energy.

If the shock is adiabatic, the density increases by a factor of 4 (equation (6.54) and therefore the magnetic field strength increases by the same factor (equation (6.87). We state without proof that the magnetic pressure of a field of strength B is proportional to B^2. Therefore, across the shock, the magnetic pressure increases by a factor of 16.

To see what would happen if the shocks were radiative, we take the special case of an isothermal shock (Section 6.4.3). Equations (6.80) and (6.87) then show that the field strength increases by a factor of M_0^4 and the magnetic pressure by a factor of M_0^2. This increase can be so large that the magnetic pressure in the shocked gas can dominate the gas pressure in the shocked gas. In this latter case, the compression of the gas can be much less than predicted by using the standard Rankine–Hugoniot conditions. Essentially what happens is that the downstream magnetic pressure more or less balances the ram pressure ($\approx \rho_0 V_s^2$) of the gas entering the shock wave.

We shall explore in the next chapter the interaction of massive stars with their interstellar environments, in which stellar winds with velocities of thousands of km s^{-1} can generate post-shock regions with temperatures of millions of degrees K or greater. In Section 6.5.2, we shall consider the effect of shocks on the largely neutral and often molecular pre-shock gas contained in interstellar diffuse and molecular clouds. The range of possible post-shock temperatures is very large, but interesting chemical effects arise in which the post-shock temperatures are low enough that the bulk of the gas remains neutral. For this to be the case, post-shock temperatures up to some thousands K and shock speeds of up to about tens of km s^{-1} may be considered.

6.5.2 Shocks in Diffuse and Dark Interstellar Clouds

In some interstellar regions, such as diffuse clouds and molecular clouds (see Section 2.5), the fractional ionization is low. In diffuse clouds, most of the ionization is maintained by the photoionization of atomic carbon, so that the fractional ionization in those regions is on the order of 10^{-4}. In dark molecular clouds where starlight is effectively excluded by dust, the ionization is maintained by the flux of cosmic rays (see Section 2.5) at a level of around 10^{-7} in clouds of number density ~10^{10} H_2 m^{-3}. Note that in dark clouds with very low ionization fractions, the magnetic field is not 'locked in' to the neutral gas, but can drift out of the neutral gas in a process known as *ambipolar diffusion*. We shall refer again to this process in Chapter 8.

The gas in these regions may be thought of as two separate fluids occupying the same volume of space. The neutral fluid is composed of the neutral atoms and molecules, while the ions and electrons comprise a separate fluid, a plasma, whose particles are directly influenced by the magnetic field; they experience Lorentz forces when they move across the magnetic field lines. The neutral particles, however, are affected by the magnetic field only through ion-neutral scattering. The detailed discussion of multi-fluid magnetic shocks is beyond the scope of this book, but we can gain some qualitative understanding from the following simple ideas.

Disturbances in the magnetic field travel at the *magnetosonic* speed, which is proportional to the magnetic field strength. If the shock speed is less than the magnetosonic speed, then disturbances in the plasma can travel upstream, ahead of the shock. These disturbances are damped by ion-neutral collisions, so that neutral gas ahead of the shock will have been affected by disturbances behind the shock by a so-called *magnetic precursor*, and this has occurred *before* the shock in the neutral gas has actually arrived.

For sufficiently weak magnetic fields, the magnetic precursor affects the neutral gas to a minor extent, and an abrupt change (a *jump*) in the velocity of the neutral gas may still occur when the shock arrives, just as in the non-magnetic case, although this abrupt change in velocity may be reduced compared to the non-magnetic case. By contrast, the velocity of the ionized gas changes continuously around the position at which the neutral gas velocity is abruptly shocked. However, for stronger magnetic fields (and therefore larger magnetosonic speeds), the velocity of the neutral gas may be so strongly affected by the magnetic precursor that the jump in the neutral velocity is entirely smoothed out by the magnetic precursor and the velocities of both the neutral and the ionized components change *continuously* with position. These distinct types of shock are called jump (J) and continuous (C) shocks (see Figure 6.9). The energy transfer in a C-type shock occurs over a larger distance, so the peak post-shock temperature in a C-type shock is lower than the peak post-shock temperature in the corresponding J-type shock of the same shock velocity.

Shocks arise in a variety of ways in the interstellar medium. Figure 6.10 shows an interesting situation involving stellar motion and a stellar wind impacting on interstellar gas.

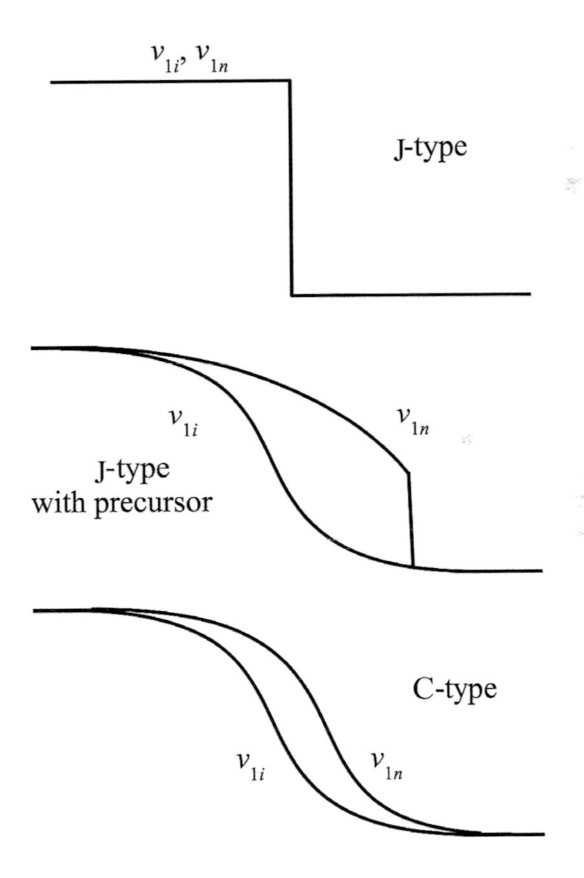

FIGURE 6.9 A schematic diagram illustrating the velocity structures of the neutral and ionized components in J- and C-type shocks in partially ionized gas, with reference to frames of reference co-moving with the shocks; upstream medium is to the left, and the material is flowing to the right. Top panel: all components are decelerated together in a J-type shock. Middle panel: an intermediate case in which the charged fluid is decelerated smoothly but the neutral fluid is affected by the precursor and shows both jump and continuous decelerations. Bottom panel: a C-type shock; both charged and neutral fluids are continuously decelerated, but not at the same rate. (Credit for figure: C. Cecchi-Pestellini.)

FIGURE 6.10 The bright 20 solar mass star (zeta Ophiuchi) at the centre of this false-colour infrared image is moving leftwards in this image at a velocity of 24 km s^{-1}. The star's powerful wind impacts on the local interstellar medium and a bow shock is created. The shock-heated post-shock medium then radiates. (Credit: NASA, JPL-Caltech, Spitzer Space Telescope.)

6.6 SOME CHEMICAL AND STRUCTURAL IMPLICATIONS OF LOW-VELOCITY SHOCKS IN INTERSTELLAR CLOUDS

The passage of a shock may affect the fractional ionization in the shocked region. In diffuse clouds, as stated in the section above, atomic carbon is photoionized by starlight and makes the major contribution to the ionization fraction. The ion–molecule exchange reaction

$$C^+ + H_2 \rightarrow CH^+ + H \tag{6.88}$$

is endothermic by about 0.4 eV and entirely suppressed at the low temperatures (~100 K) in diffuse clouds, but becomes rapid when the temperature is more than ~10^3 K, easily obtainable in low-velocity post-shock gas. The CH^+ ion reacts very quickly with H_2 in successive hydrogen abstraction reactions to form CH_2^+ and CH_3^+ ions which are then removed – also very quickly – in dissociative recombination reactions with electrons, such as

$$CH_3^+ + e \rightarrow CH + H_2 \text{ and } CH_2 + H, \quad (6.89)$$

to form the neutral molecules CH and CH_2. Reactions of this type are very much faster than radiative recombination of C^+ ions with electrons. Thus, electrons are removed quickly in this sequence of reactions and as a result the fractional ionization is reduced, by a factor of about 10.

In dark molecular clouds, the ionization is maintained by cosmic rays which ionize the most abundant species, H_2, to create H_2^+ molecular ions, which then abstract H atoms in reactions with other H_2 molecules to create H_3^+ ions. As we have seen (see Section 3.4.3) these ions stimulate a rapid chemistry by donating protons to neutral molecules such as CO and H_2O to form the molecular ions HCO^+ and H_3O^+ which are rapidly removed in dissociative recombination reactions. The H_3^+ ions also react with metal atoms such as magnesium to form Mg^+ (and other) ions in a dissociative charge exchange reaction

$$H_3^+ + Mg \rightarrow H + H_2 + Mg^+ \quad (6.90)$$

and these ions are preferentially removed by recombination on dust grains, which – because of the high thermal speeds of electrons compared to that of ions – become negatively charged in dark clouds, each grain carrying many electrons. Thus, in dark clouds, dust grains are the main carriers of charge, and the accumulation of electrons by grains increases with temperature. These charged dust grains are in effect a third fluid, beside the first fluid (neutral atomic and molecular gas) and the second fluid (ionic and electronic plasma), that should be considered in the discussion of interstellar shocks. For present purposes, it is clear that the fractional ionization is suppressed in dark clouds, that dust grains are the main carriers, and that this charging process is enhanced in the higher temperatures of shocked regions in dark clouds.

The raised temperature in interstellar shocked regions of molecular clouds enables the formation of some hydride species through reactions of atoms with molecular hydrogen, reactions that are endothermic at the normally low temperatures in quiescent gas (~10–100 K). For example, the formation of OH and H_2O (as well as CH and CH_2) can occur if the temperature is raised towards 1000 K, while the formation of SH and H_2S requires the temperature to exceed 1000 K. Temperatures in these ranges are readily achieved in shocks with velocities of a few km s^{-1}.

The response of the most abundant molecule in molecular clouds, H_2, to shocks may have important consequences for the physics and chemistry of the post-shock gas. Collisions between H_2 molecules in warm post-shock gas populate excited rovibrational levels of the molecules, and these levels are slowly radiatively depopulated, cooling the gas. If the collisional rate is high enough, then most of the hydrogen molecules will be collisionally dissociated. The critical post-shock temperature in non-magnetic dense interstellar gas is attained by J shocks with speeds of ~24 km s^{-1} while in the magnetized case the C shock speed is higher, ~45 km s^{-1}, because the peak post-shock temperature is lower behind C shocks than J shocks. Both of these shock speeds are density dependent. The consequence of creating a large amount of hot post-shock atomic hydrogen is that much of the pre-shock chemistry is destroyed. Even the highly stable molecule carbon monoxide can be destroyed in the endothermic reaction

$$\mathrm{H + CO \rightarrow C + OH}$$

at post-shock temperatures above a few thousand K, and OH molecules are similarly reduced to O atoms. Hot atomic hydrogen is a deadly stuff!

As we have seen earlier (subsection 4.3.2), dust grains may be eroded and ultimately destroyed in high-temperature gas, through sputtering by collisions with atoms and ions and through shattering in grain–grain collisions that occur because the charged grains are coupled to the magnetic field lines. Shock speeds of a few km s^{-1} are sufficiently destructive to remove grains composed of hydrocarbons and silicates, while shock speeds of around 100 km s^{-1} are necessary to remove very robust dust materials such as carbides or diamonds.

PROBLEMS

1. Estimate the mean free path for atom–atom collisions in a typical diffuse interstellar cloud. Can the internal motions in such clouds be investigated by continuum hydrodynamics?
2. Estimate the sound-crossing time of an HII region of density 10^8 m^{-3} excited by a star producing 5 $\times 10^{48}$ UV photons s^{-1}.
3. Show that solutions of the form $\rho_1 = \rho_1(x \pm a_0 t)$ satisfy equation (6.25), and that they correspond to waves propagating in the directions x increasing or decreasing.
4. Find the density jump across a strong shock propagating into a gas of constant ratio of specific heats γ = 7/5. Assume that the specific total energy of the gas is $\frac{1}{2}u^2 + [\gamma/\gamma - 1](P/\rho)$.
5. By use of the volume cooling rate given in equation (5.34) (with $y_{O+} = 1$) for an assumed temperature of 10^4 K, show that in a typical HII region shocks can quite reasonably be treated as isothermal discontinuities.
6. A strong shock wave travels at a velocity of 500 km s^{-1} through stationary interstellar gas of number density n_0 = 10^7 m^{-3} and temperature T = 10^4 K. Calculate the gas density and temperature immediately behind the shock wave. What is the gas velocity there in the shock frame and in that of the stationary interstellar gas? What is the gas velocity in these two reference frames once it has cooled back down to 10^4 K?

7. A strong shock moves with velocity 100 km s^{-1} into atomic hydrogen of density 10^8 m^{-3}. The shocked gas cools at a rate $L = 4 \times 10^{-37} n_s^2$ W m^{-3}, where n_s is the post-shock density. Estimate the time for the shocked gas to cool to a low temperature and the distance the gas has travelled relative to the shock in doing so.

7 Gas Dynamical Effects of Stars on the Interstellar Medium

7.1 EXPANSION OF IONIZED NEBULAE AROUND MASSIVE STARS

7.1.1 INTRODUCTION

We showed in Chapter 5 that a massive hot star will photoionize the interstellar gas in its neighbourhood. In doing so, it increases the gas temperature from about 10^2 K to about 10^4 K, that is, by a factor of about a hundred. The ionization process itself increases the number of gas particles, and therefore the pressure, by a further factor of two. As a result, the pressure in the ionized gas is two hundred times greater than that in surrounding neutral material. This ionized gas cannot be confined and will expand. Both it and the adjacent neutral gas are set in motion, so that the HII regions described in Chapter 5 are not static configurations.

This particular process is not the only way in which interstellar gas is set in motion by means of interaction with stars. Two other mechanisms are of particular importance. Firstly, there are the effects produced by the very high speed continuous mass loss – a stellar wind – observed to take place from many young hot stars. Secondly, many massive stars terminate their existence in a violent explosive event – a supernova. The effects of these explosions on the interstellar gas are extremely important in determining many of its properties.

We shall discuss in this chapter all three of these processes using simple models which, although approximate, give quite good descriptions of the events taking place. They serve to show how the ideas of gas dynamics discussed in Chapter 6 can be put into practice. More sophisticated treatments must, however, be done numerically because of the inherently complicated interplay of various physical effects.

7.1.2 A SIMPLE MODEL FOR THE EXPANSION OF A PHOTOIONIZED NEBULA

The simplest possible model of a photoionized nebula is one in which a single massive young star is born somewhere inside a uniform cloud of atomic hydrogen of unlimited extent. Any disturbances to the cloud structure produced by the formation of the star are neglected. After a relatively short time ($\lesssim 10^5$ yr), the star reaches a static configuration in which it can remain for a much longer time ($\gtrsim 3 \times 10^6$ yr). The stellar radiant energy output rate and the spectral distribution of the radiation are more or less constant during this phase. The star produces Lyman continuum photons at a constant rate (S_*). Since the preceding phase is relatively short, we shall ignore it and take the star to be 'switched on' instantaneously. A sphere of ionized gas is formed around the star. As we shall see shortly, the boundary between the ionized and neutral material moves into the surroundings at a speed which generally is much less than the speed of light. The radius of the ionized gas sphere is a function of time. It was demonstrated in Chapter 5 that the ionized–neutral boundary is very sharp in the purely static case. This conclusion is generally valid even when the boundary moves but, in this case, it is usually called an ionization front (I-F for short). The sharpness of the boundary has the very important consequence that the I-F has a thickness much less than the typical dimensions of the ionized region and its properties can be determined by physical considerations applied using plane parallel geometry, in a manner analogous to the derivation of the Rankine–Hugoniot conditions, which hold across shock fronts.

7.1.3 The Velocity of the Ionization Front

A basic property of an I-F is the velocity at which it moves. It is important to define clearly the frame of reference in which this velocity is measured. If measured with respect to the star, this is a fixed frame of reference. Alternatively, we know that as the I-F moves outwards from the star, it ionizes progressively more and more gas and must therefore have a finite over-taking velocity measured with respect to the neutral material. Of course, it also has a finite velocity with respect to the ionized material its passage produces.

Let the gas around the star be at rest (in the frame of reference of the star). At time t, the I-F is a distance R from the star and at time $t + \mathrm{d}t$, it is at a distance $R + \mathrm{d}R$. A small section of the I-F is shown in Figure 7.1(a).

Let the undisturbed neutral hydrogen number density be n_0 m^{-3} and let J m^{-2} s^{-1} be the number of Lyman continuum photons falling normally on unit area of the I-F per second. (Later on we relate J to S_*.) Everywhere interior to the I-F, the gas is effectively fully ionized (as we saw in Chapter 5). Therefore, in moving the I-F from R to $R + \mathrm{d}R$, enough photons must have arrived at the front to ionize all the neutral atoms lying between these two positions. For unit area of I-F, the relationship

$$J\mathrm{d}t = n_0\,\mathrm{d}R \tag{7.1}$$

must be satisfied. This can be written as

$$\frac{\mathrm{d}R}{\mathrm{d}t} = \frac{J}{n_0}\,. \tag{7.2}$$

The velocity given by this expression is the velocity with which the I-F moves relative to the neutral gas. Since, however, it was earlier assumed that this gas

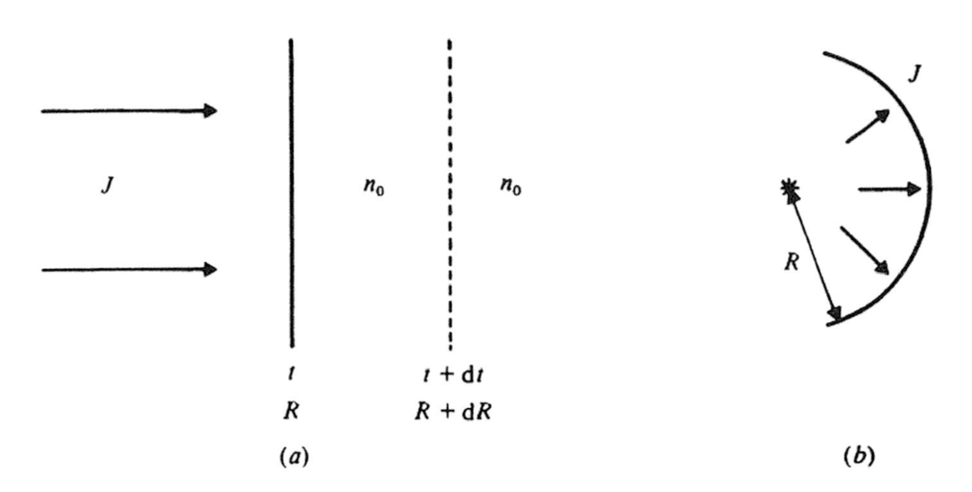

FIGURE 7.1 Geometry of an ionization front (I-F). (a) A small part of the I-F is treated as planar. (b) The I-F is spherical.

is at rest, it is, also in this case, the velocity with which the I-F moves in a fixed frame of reference. Implicit in this derivation is the assumption that only one photon is needed to ionize each atom as the front overruns the neutral gas. Or, alternatively stated, no recombination occurs within the front. Generally, this is true for all except very slow fronts.

7.1.4 The Radius of the Ionization Front

In order to relate J to S_*, two effects must be considered. Firstly, the I-F is really a spherical surface centred on the star (see Figure 7.1(b)) and the radiation field at the I-F is diluted because of the spherical geometry. Secondly, recombination takes place continuously inside the ionized region and neutral atoms are continuously being created. These atoms absorb photons travelling outwards from the star and cause further reduction in the flux at the I-F.

J and S_* can be related just by counting photons using the requirement that the photon output rate from the star must equal the rate at which they arrive at the I-F plus the rate at which they are absorbed by neutral atoms. If we assume, as in Chapter 5, that ionization balance holds, then this condition can be written as

$$S_* = 4\pi R^2 J + \frac{4}{3}\pi R^3 n_0^2 \beta_2. \tag{7.3}$$

Hence, J is given by

$$J = \frac{S_*}{4\pi R^2} - \frac{1}{3} R n_0^2 \beta_2. \tag{7.4}$$

Two further comments are necessary here. The ionized gas density is assumed equal to the neutral gas density n_0. This is justified qualitatively in Section 7.1.5. Equation (7.3) is an expression of conservation for the stellar photons. The diffuse UV photons produced by ground state recombinations of hydrogen are taken into account using the 'on-the-spot' approximation (see Section 5.2.4). The recombination coefficient β_2 is used, appropriate for excited level recombinations only.

Equations (7.2) and (7.4) give for the I–F velocity

$$\frac{\mathrm{d}R}{\mathrm{d}t} = \frac{S_*}{4\pi R^2 n_0} - \frac{1}{3} R n_0 \beta_2. \tag{7.5}$$

Obviously, $\mathrm{d}R/\mathrm{d}t$ decreases with increasing R. If the Strömgren radius, R_S, is defined as that radius at which the stellar photon output rate just balances the recombination rate in the entire ionized volume, then from equation (5.15)

$$R_\mathrm{S} = \left(\frac{3}{4\pi}\frac{S_*}{n_0^2\beta_2}\right)^{1/3}. \tag{7.6}$$

The characteristic time, t_R, in which hydrogen recombines is

$$t_\mathrm{R} \approx (n_0\beta_2)^{-1}. \tag{7.7}$$

Using equations (7.6) and (7.7), we may define the following dimensionless quantities:

$$\begin{aligned} \lambda &= R/R_\mathrm{S} \qquad & V_\mathrm{R} &= R_\mathrm{S}/t_\mathrm{R}. \\ \tau &= t/t_\mathrm{R} \qquad & \dot{\lambda} &= (\mathrm{d}R/\mathrm{d}t)/V_\mathrm{R}. \end{aligned} \tag{7.8}$$

Equation (7.5) in dimensionless form becomes

$$\dot{\lambda} = \frac{1}{3}(1 - \lambda^3)/\lambda^2, \tag{7.9}$$

which has as its solution

$$\lambda = (1 - \mathrm{e}^{-\tau})^{1/3}. \tag{7.10}$$

This implies $\lambda \to 0$ as $\tau \to 0$. Values of τ and corresponding values of λ and $\dot{\lambda}$ are shown in Table 7.1. For illustrative purposes throughout this chapter, values of 10^{49} s^{-1} and 2×10^{-19} m^3 s^{-1} will be used for $S_*$and β_2, respectively (as in Chapter 5). Then, t_R and R_S have characteristic values of $5 \times 10^{18} n_0^{-1}$ s and $2 \times 10^{22} n_0^{-2/3}$ m, respectively. The associated value of V_R is $4 n_0^{1/3}$ km s^{-1}. Table 7.1 shows also numerical values of $\mathrm{d}R/\mathrm{d}t$ as a function of time, using $n_0 = 10^8$ m^{-3}.

TABLE 7.1
Evolution of the I-F

τ	$\dot{\lambda}$	Λ	dR/dt (km s^{-1})
0.1	1.42	0.46	2630
0.7	0.25	0.80	460
0.9	0.19	0.84	352
1.0	0.16	0.86	300
2.0	0.05	0.95	93
4.0	0.006	0.99	11

7.1.5 Results of the Model

Several very important conclusions can be drawn from Table 7.1.

(i) The Strömgren radius, which is defined using the density of the interstellar gas around the star, is reached only after a large time. However, R is within a few per cent of R_S at times greater than about t_R.
(ii) Until R is very close to R_S, the radius of the ionized region increases at a speed much greater than the characteristic speed (the sound speed $c_i \approx 10$ km s^{-1}) at which the ionized gas reacts to the sudden pressure increase caused by photoionization. The ionized gas cannot move appreciably during this phase and the gas density consequently remains unchanged by the ionization process. This is the reason why the ionized and neutral gas densities can be assumed equal.
(iii) When a time is reached greater than a few times t_R, the I-F slows down very rapidly. Eventually this simple model would predict an I-F velocity equal to and then below the sound speed c_i in the ionized gas. However, as mentioned in Section 7.1.1, the pressure in the ionized gas is much greater than the surrounding matter. It tries to expand with a velocity of about c_i. Obviously, then the I-F must sit around the expanding sphere of ionized gas and the motion of the I-F is coupled to the expansion of the sphere.

7.1.6 The Final Stage of Evolution of a Photoionized Region

Before analyzing the evolution of the HII region in the stage $t \gtrsim t_R$, it is instructive to consider the ultimate state which could be reached by an HII region. The ionized gas expands as long as it has a higher pressure than its surroundings. This expansion, in principle, will cease when the hot ionized gas reaches pressure equilibrium with the surrounding cool neutral gas.

The condition of final pressure equilibrium can be stated in the form

$$2n_f kT_i = n_0 kT_n. \tag{7.11}$$

n_f is the ionized gas density at this stage. T_i and T_n are, respectively, the ionized and neutral gas temperature. As representative values, we take $T_i = 10^4$ K and $T_n = 100$ K. (The factor 2 on the left-hand side of equation (7.11) allows for the fact that the number of particles doubles on photoionization.) The ionized gas sphere must still absorb all the stellar UV photons. Thus,

$$S_* = \frac{4}{3}\pi R_f^3 n_f^2 \beta_2. \tag{7.12}$$

R_f is the final radius of the ionized gas sphere. From equations (7.6), (7.11), and (7.12), therefore,

$$n_f = (T_n/2T_i)n_0 \tag{7.13}$$

and

$$R_f = (2T_i/T_n)^{2/3} R_S. \tag{7.14}$$

The ratio of the mass of gas finally ionized (M_f) to that (M_S) contained within the initial Strömgren sphere of radius R_S is

$$\frac{M_f}{M_S} \equiv \frac{R_f^3 n_f}{R_S^3 n_0} = \frac{2T_i}{T_n}. \tag{7.15}$$

Thus, typically $n_f/n_0 \approx 0.005$, $R_f/R_S \approx 34$, $M_f/M_S \approx 200$. The initial Strömgren sphere contains only a very small fraction of the material which, in principle, a star could ultimately ionize.

The density of the ionized gas in the final configuration is much less than the density of the gas which was originally contained within the sphere of radius R_f. Evidently, neutral gas must be displaced from its original position during the expansion of the ionized sphere. Now, in Chapter 6, it was demonstrated that a shock wave is set up when a piston is advanced supersonically into a gas. The expansion velocity of the ionized gas sphere (which is bounded by the I-F) is originally equal to about c_i. This is highly supersonic with respect to the sound speed c_n ($c_i/c_n \lesssim 10$) in the neutral gas. Therefore, the ionized gas sphere plays the role of a piston and pushes a shock wave into the neutral gas. The shock then sets the neutral gas into motion outwards and it is therefore displaced from its original position.

It should be noted that the velocity of the I-F relative to the neutral gas (equation (7.2) would be expected to drop when the shock wave is formed. The shock wave is compressive and its passage increases the neutral density ahead of the I-F to a value greater than n_0. However, in the frame of reference of the star, the I-F velocity is really fixed by the expansion velocity of the ionized gas sphere around which the front sits. This can be much greater than its velocity relative to the neutral gas, and so in the stellar frame, the I-F velocity hardly changes.

7.1.7 Assumptions Concerning the Intermediate Stages of Evolution of the Ionized Region

A simplified scheme for the evolution of an HII region from the initial stages, described in Section 7.1.4, to the final pressure equilibrium stage of Section 7.1.6 can now be made. It is sketched in Figure 7.2(a)−(c).

In the first stage, $R \lesssim R_S$ and the motion is as described by equations (7.9) or (7.10). The ionized gas is at rest with the same density (n_0 m^{-3}) as the surroundings. In the second phase ($R \gtrsim R_S$), a shock wave driven by the expanding bubble of hot gas moves directly into the neutral material and sets it into outwards motion. Several simplifying assumptions must be made in order to proceed with the analysis. We now list them with appropriate comments.

(i) The layer of shocked neutral gas is thin because of the compression across the shock. A single radius R and velocity dR/dt then can refer to both shock and I-F. (Note that it is mentioned earlier that the I-F velocity with respect to the neutral gas is low compared with its velocity in a fixed frame.)

(ii) The pressure behind the shock wave is assumed uniform (but time dependent) in both the neutral and ionized gas. For this to be true, the sound travel times across the neutral and ionized regions should be less than the timescale for a substantial change in the overall configuration. This is a very good approximation in the thin neutral

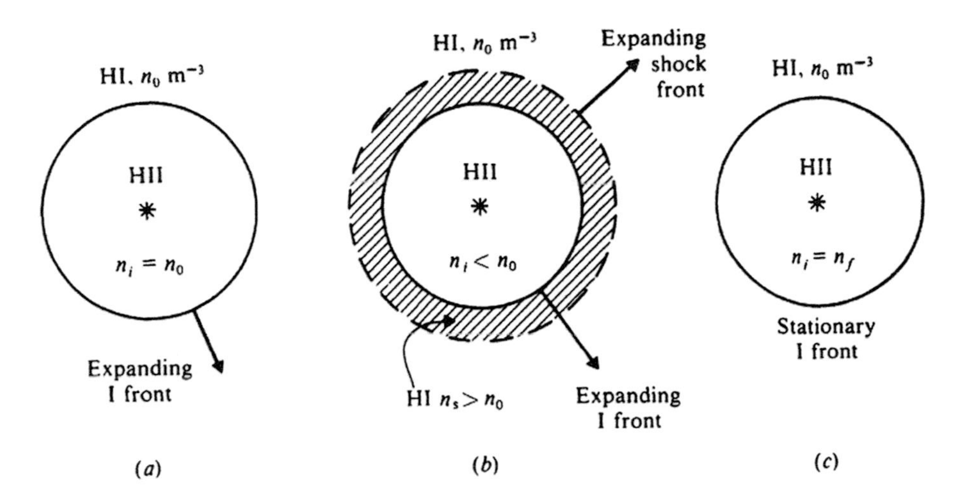

FIGURE 7.2 Evolutionary scheme of an expanding HII region: (a) the initial stages, (b) expansion with a shock in the neutral gas, and (c) the final equilibrium state.

layer. It is not such a good approximation in the ionized gas, at least in the early stages of expansion, where the ionized gas sphere expands with a velocity comparable to c_i. It gets progressively better and must in any case be used in order to make a simple analysis. Since the temperature in the ionized gas varies little (see Chapter 5), this assumption implies a uniform density (n_i m^{-3}) in the ionized region.

(iii) The shock is strong, that is, it moves highly supersonically into the neutral material. This then enables us to use the strong-shock relationships derived in Chapter 6. Of course, the shock weakens with time, and this assumption eventually breaks down.

(iv) The recombination rate in the ionized gas balances the stellar UV output rate. Strictly, this can never be satisfied exactly, as a fraction of the stellar photon output must be used up in ionizing fresh gas since the ionized mass is always increasing with time. It can, however, be demonstrated that this fraction is small.

(v) The neutral gas ahead of the shock is at rest. Then, dR/dt is the shock velocity both relative to the neutral gas and in the stellar frame of reference.

7.1.8 Intermediate Evolution: The Model

We now put assumptions (i)–(v) into a simple model. The pressure behind the shock wave, P_s, is equal to the pressure P_i in the ionized gas given by

$$P_i = 2n_i kT_e \equiv n_i m_H c_i^2. \tag{7.16}$$

In Chapter 6 (Sections 6.3.4 and 6.3.5), it was shown that the pressure, P_s, behind a strong shock is related to the shock velocity, V_s (relative to the upstream gas), and upstream density ρ_0 by

$$P_s = \varepsilon \rho_0 V_s^2, \tag{7.17}$$

where $\varepsilon = \frac{3}{4}$ if the shock is adiabatic, and $\varepsilon = 1$ if the shock is isothermal. The cooling in the shocked neutral gas can be extremely effective by processes such as those outlined in Chapter 3. Therefore, as a reasonable assumption, we take $\varepsilon = 1$. Equation (7.17) then can be written as

$$P_s = P_i = n_0 m_H \dot{R}^2, \tag{7.18}$$

where $\dot{R} = \mathrm{d}R/\mathrm{d}t \equiv V_s$. Thus, from equations (7.16) and (7.18),

$$\dot{R}^2 = (n_i/n_0)c_i^2. \tag{7.19}$$

Assumption (iv) of Section 7.1.7 implies that

$$S_* = \frac{4}{3}\pi n_i^2 \beta^2 R^2. \tag{7.20}$$

Using the previous definition of R_s (equation (7.6) and equations (7.19) and (7.20), we obtain

$$R^{3/2}\dot{R}^2 = c_i^2\left(\frac{3S_*}{4\pi\beta_2 n_0^2}\right)^{1/2} \equiv c_i^2 R_s^{3/2}. \tag{7.21}$$

It is again convenient to use dimensionless variables, but now in the following forms:

$$\lambda = R/R_s \quad N = c_i t/R_s\,. \tag{7.22}$$

Equation (7.21) then takes the very simple form

$$\lambda^{3/4}\mathrm{d}\lambda/\mathrm{d}N = 1. \tag{7.23}$$

In order to find a suitable boundary condition for solving equation (7.23), we will assume that the time taken to set up the initial Strömgren sphere is a very small fraction of the lifetime of the HII region, that is, $N = 0$ at $\lambda = 1$. This will be justified later. Using this condition, the solution of equation (7.23) is

$$\lambda = \left(1 + \frac{7}{4}N\right)^{4/7}. \tag{7.24}$$

or

$$\mathrm{d}\lambda/\mathrm{d}N \equiv \dot{\lambda} = \left(1 + \frac{7}{4}N\right)^{-3/7}. \tag{7.25}$$

Note that equation (7.25) implies that $\dot{R} = c_i$ at $N = 0$, which is consistent with the qualitative discussion in Section 7.1.6. Table 7.2 gives values of λ and $\dot{\lambda}$ as functions of N. Obviously, the expansion speed of the sphere drops quickly below c_i.

We are now in a position to verify our assumption that the initial Strömgren sphere is set up in a timescale much less than the time, t_e, spent in the expansion phase. From Table 7.2, $t_e \gtrsim R_s/c_i$; then $t_e/t_R \,\tilde{>}\, 0.4n_0^{1/3}$ using the characteristic values of R_s and t_R given in Section 7.1.4. Thus, $t_e/t_R \gg 1$ as required.

7.1.9 Does the Strömgren Sphere Reach Pressure Equilibrium?

Further, we can use the previous result to estimate the time required, t_{eq}, to set up the final pressure equilibrium Strömgren sphere. From Section 7.1.6, the value

TABLE 7.2

N	λ	$\dot{\lambda}$	$\lambda^3\dot{\lambda}^2/N$
0.1	1.10	0.93	11.5
0.5	1.43	0.76	3.43
1	1.78	0.65	2.38
5	3.67	0.38	1.41
20	7.75	0.22	1.08
100	19.2	0.11	0.84
250	32.3	0.074	0.74
300	35.9	0.068	0.71

of λ ($\equiv R/R_S$) at this stage is approximately 34. Equation (7.24) then gives the corresponding value of N as about 273; hence, $t_{eq} \approx 273\ R_s/c_i$. With the characteristic value of R_s used before, then $t_{eq} \approx 1.7 \times 10^{13} n_0^{-2/3}$ yr. For example, with $n_0 = 10^8\ m^{-3}$, $t_{eq} = 7.8 \times 10^7$ yr. This is typically an order of magnitude greater than the constant-luminosity (or main-sequence) phase of existence of a massive star. We conclude that unless the interstellar gas density around a star is extremely high to start with ($n_0 \gtrsim 3 \times 10^9\ m^{-3}$), it is unlikely that the final pressure equilibrium configuration can be reached. This conclusion is further strengthened by the fact that the above estimate for t_{eq} will be too low. For the strong-shock approximation to be valid, $\dot{R}$ must be much greater than c_n ($\approx c_i/10$ typically). But we note from Table 7.2 that $\dot{R} \approx c_i/10$ when $t \approx 100R_s/c_i$, that is, before $\lambda \approx 34$. Hence, the expansion rate in the very last stages is overestimated by this treatment.

7.1.10 Conversion Efficiency for Stellar UV to Kinetic Energy

The kinetic energy of the gas during this expansion phase must ultimately derive from the energy of the stellar UV radiation field. It is interesting to investigate quantitatively the efficiency of conversion of this radiant energy into kinetic energy.

At time t the total UV energy which has been absorbed by the gas is

$$E_* = S_* \langle h\nu \rangle t \ . \tag{7.26}$$

In equation (7.26), $\langle h\nu \rangle$ is the average energy of a stellar photon. Energy I_H ($\equiv$ 13.6 eV) is immediately lost from each photon due to the process of photoionization. For our order of magnitude estimate, we can neglect this. The expansion velocity of the ionized gas is generally much less than the sound speed, c_i (see Table 7.2). Since the kinetic energy is proportional to the expansion velocity squared, and the thermal energy to the sound speed squared, the energy of the ionized gas is mainly in the form of thermal energy. If this energy is E^i_{th}, then

$$E_{\rm th}^{\rm i} \approx \left(\frac{4}{3}\pi R^3\right)(2n_{\rm i})\left(\frac{3}{2}kT_{\rm e}\right) \approx 2\pi R^3 n_{\rm i} m_{\rm H} c_{\rm i}^2. \tag{7.27}$$

The opposite is true for the neutral gas, which is assumed to have cooled back to its original temperature. The energy is mainly kinetic and is given by

$$E_{\rm K}^{\rm n} = \frac{1}{2}\left(\frac{4}{3}\pi R^3 n_0 m_{\rm H} - \frac{4}{3}\pi R^3 n_{\rm i} m_{\rm H}\right)\dot{R}^2. \tag{7.28}$$

In the later stages, $n_i \ll n_0$ and hence,

$$E_{\rm K}^{\rm n} \approx \frac{2}{3}\pi R^3 n_0 m_{\rm H} \dot{R}^2. \tag{7.29}$$

Hence,

$$\frac{E_{\rm K}^{\rm n}}{E_{\rm th}^{\rm i}} \approx \frac{1}{3}\frac{n_0}{n_{\rm i}}\frac{\dot{R}^2}{c_{\rm i}^2} \approx \frac{1}{3}$$

from equation (7.19), that is, they are of comparable magnitude. The fraction, f, of the stellar energy output which has been converted into kinetic energy of surrounding gas is therefore approximately

$$f \approx \frac{E_{\rm K}^{\rm n}}{E_*} \approx \frac{2\pi R^3}{3}\frac{n_0 m_{\rm H} \dot{R}^2}{S_* \langle h\nu \rangle t}. \tag{7.30}$$

Using equations (7.6), (7.19), (7.22), and (7.23), equation (7.30) can be written as

$$f \approx \frac{m_{\rm H} c_{\rm i}^3}{2\langle h\nu \rangle \beta_2 n_0} \frac{1}{R_{\rm s}} \left(\frac{\lambda^3 \dot{\lambda}^2}{N}\right). \tag{7.31}$$

As a typical value of $\langle h\nu \rangle$, we take 2×10^{-18} J along with the previously used values of $c_{\rm i}$, β_2, and $R_{\rm s}$. Thus,

$$f \approx 10^{-1} n_0^{-1/3} \lambda^3 \dot{\lambda}^2 / N. \tag{7.32}$$

Values of the function $\lambda^3 \dot{\lambda}^2 / N$are given in Table 7.2. Quite evidently, less than 1% of the stellar UV energy is converted into kinetic (or thermal) energy of the surrounding gas. Nearly all the incident UV radiation is radiated away by the forbidden-line processes discussed in Section 5.3.

7.2 THE EFFECTS OF THE STELLAR WINDS OF MASSIVE STARS ON THE INTERSTELLAR GAS

7.2.1 INTRODUCTION

Observations of the spectra of the very luminous hot stars which ionize diffuse nebulae show that these stars must be losing material in a continuous stream from their surfaces. This continuous mass loss is generally referred to as a *stellar wind.* This mass loss affects the stellar luminosity and lifetime in a complex fashion, but we shall not consider such effects any further. The physical mechanism which drives the wind involves the conversion of momentum of the stellar radiation field into momentum of stellar gas via absorption by ions such as C^{3+}.

There are several important physical parameters relating to the mass loss which can be estimated from the observations and which will turn out to be relevant to our models. The wind velocity (V_*) is extremely high, characteristically about three times the escape velocity from the star, and we take $V_* \approx 2000\text{kms}^{-1}$ as typical. This is a factor of a hundred or so greater than the sound speed in the wind. Note that because the wind velocity is so high the thermal energy in the wind is negligible. We assume that the mass loss rate from the star $\dot{M}_*$ is constant in time. We adopt a value of $10^{-6}\ M_\odot\ \text{yr}^{-1}$, which is characteristic of a massive O star. This implies that the stellar mechanical energy output rate (the 'mechanical luminosity') of the wind $(\dot{E}_* \equiv \frac{1}{2}\dot{M}_* V_*^2)$ is about 10^{29} J s^{-1}. The stars have a typical UV photon output rate of about 10^{49} s^{-1} with an associated radiative energy output rate $\approx 10^{31}$ J s^{-1} in the Lyman continuum. This is two orders of magnitude greater than the mechanical energy output rate. However, we have already seen (Section 7.1.10) that the efficiency of conversion of stellar UV radiant energy into gas kinetic energy is extremely low. We cannot immediately rule out the possibility that the conversion of the wind mechanical energy into gas kinetic energy may be so efficient that it compensates for the difference in energy output rates. Indeed, we shall see in Section 7.2.5 that this is exactly the case.

7.2.2 THE IMPACT OF THE STELLAR WIND ON THE SURROUNDING GAS

We assume that the star is surrounded by an HII region and therefore the gas involved in the flow is fully ionized. This assumption will be checked later (Section 7.2.6). The impact of the wind on the surrounding gas produces regions of gas flow whose qualitative features can be deduced from the principles discussed in Chapter 6.

The wind (velocity $V_* \approx 2000\text{kms}^{-1}$) pushes the interstellar gas (whose sound speed is $c_{\text{i}} \approx 10$ km s^{-}) at a speed which, certainly initially, is highly supersonic. The wind acts as a piston and a shock wave must immediately be set up in the interstellar gas. In doing this, the wind itself must be slowed down. Since the wind itself has only kinetic energy, some way must be found of converting this energy into thermal energy. The only way of doing this is to introduce a second shock into the wind itself. (Alternatively, we could say that the shell of swept-up interstellar matter plays the part of an obstacle or piston introduced into the wind flow. The wind moves supersonically towards this piston, or, changing the frame of reference, the piston advances supersonically into the wind. A shock naturally results.) We should note, however, that this shock faces towards the star unlike the shock in the interstellar gas. This flow pattern is sketched in Figure 7.3. There are four distinct regions of flow and three distinct boundaries marked in the figure. The main properties of the flow regions and the roles they play in the flow will now briefly be discussed.

7.2.3 The Regions of the Flow Pattern

Region (*a*) is occupied by unshocked stellar wind gas moving with a velocity V_* ($\approx$ 2000 km s^{-1}). It enters shock S_1, which converts part of the energy of this wind into thermal energy.

Region (*b*) contains shocked stellar wind gas. Since the Mach number of this shock is very high, this shock is very strong. The immediate post-shock gas temperature is (equation (6.58))

$$T_s = \frac{3}{32}\frac{m_H V_*^2}{k} \approx 4 \times 10^7\text{K}. \tag{7.33}$$

(In using equation (7.33), it has been assumed that shock S_1 moves with a velocity $\ll V_*$ in the stellar frame of reference. This is always a good approximation.) The sound speed associated with this gas is therefore very high (approximately 600 km s^{-1}). It is the expansion of this hot bubble of gas which drives the shell of shocked interstellar gas (*region* (*c*)) and this is

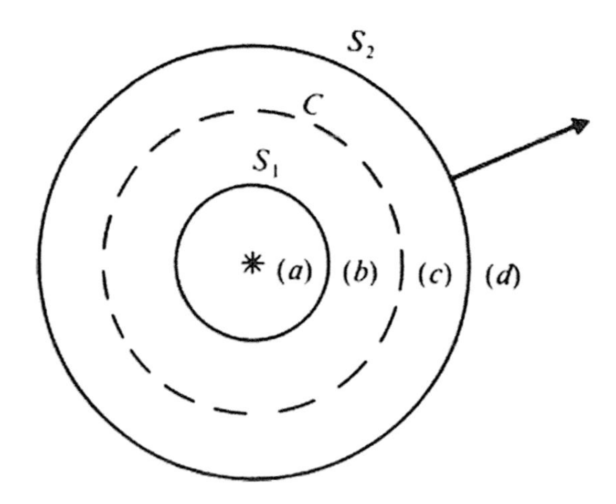

FIGURE 7.3 The interaction of a stellar wind with interstellar gas.

sometimes referred to as an 'energy driven' flow. Now, the expansion velocity of the bubble drops because of the loss of internal energy as it does work on the surrounding material. It turns out that this velocity very quickly drops to a value much less than the internal sound speed. The sound crossing time becomes much less than the expansion timescale and this gas therefore has a uniform pressure (Section 6.2), which decreases with time.

A key question is whether or not this shocked wind gas can lose energy by the radiative processes described in Section 6.4.1. Figure 6.6 shows that at a temperature of 4 x 10^7 K the radiative cooling rate is nearly at its minimum value. We saw in Section 6.4.2 that, at such a temperature, the cooling time behind a shock of velocity V_s is proportional to $V_s^3 n_0^{-1}$, where n_o is the density ahead of the shock. Here, $V_s \equiv V_*$ and n_0 is the wind density at the wind shock. Since the wind mass loss rate from the star is constant, the wind density at distance r from the star is $n_0 = \dot{M}_*/4\pi r^2 \bar{m} V_*$. Thus, the cooling time $t_c \propto V_*^4 r_s^2$, where r_s is the distance of the shock from the star. As noted, V_* is very high and characteristic values of r_s also turn out to be high. The cooling time becomes far larger than even the stellar lifetime. Thus, the total energy in this region is determined by the rate at which the gas does work on its surroundings and the rate at which energy is fed in through shock S_1. Because the expansion velocity is much less than the sound speed, the energy in this region is almost entirely thermal. Finally, because this gas does not cool easily, it is not greatly compressed (by the adiabatic factor of 4 only) and hence conservation of mass means that this region is spatially extended.

Region (*c*) is a shell of interstellar gas which has passed through the outer shock S_2. Radiative cooling is now extremely effective in this region for two reasons. Firstly, the density immediately behind shock S_2 is four times the ambient density (which is far larger than the wind density at typical values of r_s) and is not affected by geometrical dilution. Secondly, the interstellar gas enters shock S_2 at a much lower speed than the velocity at which the stellar wind gas enters shock S_1. The immediate post-shock temperature is therefore considerably lower than that behind S_1. The gas behind S_2 contains ions (such as O^{++}) which can be collisionally excited by electrons to produce the forbidden lines discussed in Section 5.3 or even to excited-state configurations which can decay back to the ground level by allowed transitions. Cooling is therefore very effective.

As we saw in the discussion on isothermal shocks in Section 6.4.3, radiative cooling drastically increases the compression behind a shock wave. Therefore, by the requirement of continuity, *region* (*c*) must be extremely thin. The cooling of the gas in this region does not, however, continue down to a very low temperature. This gas is still exposed to the UV radiation from the star emitting the wind. The gas cools down only to a temperature (approximately 10^4 K) at which the cooling rate is equal to the energy input rate due to photoionization by the stellar radiation field. Therefore, it has the properties of a typical radiatively excited HII region. Since this region is so thin, with a resulting small sound travel time across it, it is essentially at constant pressure. Thus, the entire region between shocks S_1 and S_2 has a uniform pressure.

There is a further surface of discontinuity in the flow pattern, C (Figure 7.3), which separates shocked stellar wind gas and shocked interstellar gas. Across this surface, the temperature, density, and other physical characteristics (e.g. the ionization state) of the gas change discontinuously. However, as is clear from the remarks mentioned earlier, the pressure remains unchanged. This implies that there is no mass flow across the surface. It is usually called a 'contact' or 'tangential' discontinuity.

Region (*d*) is the surrounding ionized interstellar gas. On general grounds we would expect velocities of the order of the sound speed (10 km s^{-1}) to be present. However, in our simplified analysis, shock S_2 will be assumed strong, and so to a first approximation, this gas will be assumed to be at rest.

7.2.4 A Simple Model

The qualitative arguments in the previous section can now be used to construct a simple model whose properties can easily be investigated. *Region* (*c*) will be assumed to be a very thin shell. A single radius, R, can then be used both for the radius of the thin shell and for the outer radius of the bubble of hot shocked wind gas. Further, since *region* (*b*) is thick, we will assume that the shocked wind gas occupies the entire volume interior to the shock surface of radius R. Since the interstellar gas is at rest, $\dot{R}$ ($\equiv \mathrm{d}R/\mathrm{d}t$) represents both the shock velocity relative to the interstellar gas and the expansion velocity of the bubble. This simple model is sketched in Figure 7.4. Physically, the thin shell (*region* (*c*)) is compressed and accelerated by the hot bubble which has uniform pressure, $\mathcal{P}$, say.

The equation of conservation of momentum of the shell has the form

$$\frac{\mathrm{d}}{\mathrm{d}t}\left(\frac{4}{3}\pi R^3 n_0 m_\mathrm{H} \dot{R}\right) = 4\pi R^2 \mathcal{P}. \tag{7.34}$$

The term on the left-hand side of equation (7.34) represents the rate of change of momentum of the shell, while the term on the right-hand side represents the force acting on it. Equation (7.34) can be written in the form

$$\mathcal{P} = \rho_0 \dot{R}^2 + \frac{1}{3}\rho_0 \ddot{R} R. \tag{7.35}$$

As noted in Section 7.2.3, the energy content of *region* (*b*) is almost entirely in the form of thermal energy. The thermal energy per unit volume of a monatomic gas is equal to $\frac{3}{2}$ times the gas pressure. The total thermal energy in *region* (*b*) is therefore equal to $\frac{4}{3}\pi R^3\left(\frac{3}{2}\mathcal{P}\right)$. Conservation of energy for this region demands that

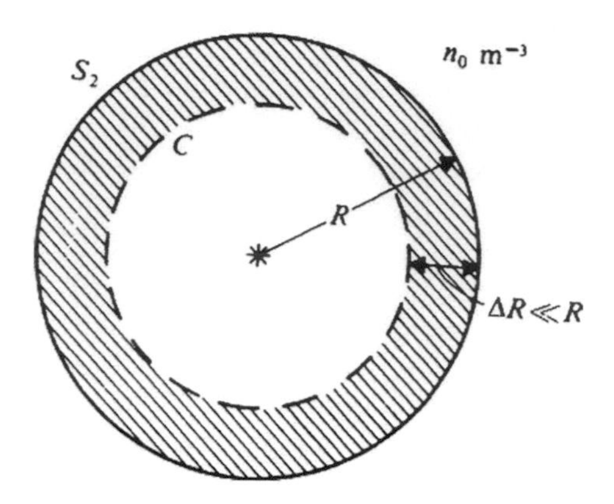

FIGURE 7.4 A simple model for the interaction of a stellar wind with interstellar gas.

$$\frac{\mathrm{d}}{\mathrm{d}t}\left[\frac{4}{3}\pi R^3\left(\frac{3}{2}\mathcal{P}\right)\right] = \dot{E}_* - \mathcal{P}\frac{\mathrm{d}}{\mathrm{d}t}\left(\frac{4}{3}\pi R^3\right). \tag{7.36}$$

Equation (7.36) states that the rate of change of thermal energy of *region* (*b*) is equal to the rate at which energy is fed into the region by gas entering shock S_1, minus the rate at which the hot gas does work on the interstellar gas. (This implicitly assumes that all the kinetic energy of the gas entering shock S_1 is immediately converted into thermal energy. This in fact cannot happen just by gas crossing the shock – see Section 6.3.5. Rather, it occurs because of the very slight pressure gradients which must actually exist in *region* (*b*) and which quickly slow the gas down to very subsonic velocities.) Equations (7.35) and (7.36) can be combined to give the equation of motion of the shell as

$$R^4\dddot{R} + 12R^3\dot{R}\ddot{R} + 15R^2\dot{R}^3 = \frac{3}{2\pi}\frac{\dot{E}_*}{\rho_0} \tag{7.37}$$

(where $\ddot{R} = \mathrm{d}\dot{R}/\mathrm{d}t$ etc.). Rather than looking for a general solution for equation (7.37), we shall pick a form of dependence of R on t and show that it both satisfies equation (7.37) and has a reasonable behaviour in the initial stages. We shall assume that R is given by an expression of the form

$$R = At^{\alpha}, \tag{7.38}$$

where A and α are positive constants. R then increases with time and tends to zero at t = 0, as seems reasonable. Substitution of equation (7.38) and its derivatives into equation (7.37) then leads to the relationship

$$A^5\left[\alpha(\alpha-1)(\alpha-2)+12\alpha^2(\alpha-1)+15\alpha^3\right]t^{5\alpha-3}=\frac{3\dot{E}_*}{2\pi\rho_0}. \tag{7.39}$$

Since the mechanical energy output rate of the star is assumed constant in time, both the right-hand and left-hand sides of equation (7.39) are time independent and hence $\alpha=\frac{3}{5}$. It immediately follows that

$$A=\left(\frac{125}{154\pi}\right)^{1/5}\left(\frac{\dot{E}_*}{\rho_0}\right)^{1/5}. \tag{7.40}$$

Therefore, we find that

$$R=\left(\frac{125}{154\pi}\right)^{1/5}\left(\frac{\dot{E}_*}{\rho_0}\right)^{1/5}t^{3/5} \tag{7.41}$$

and

$$\dot{R}=\frac{3}{5}\left(\frac{125}{154\pi}\right)^{1/5}\left(\frac{\dot{E}_*}{\rho_0}\right)^{1/5}t^{-2/5}. \tag{7.42}$$

R and $\dot{R}$ are related by

$$R=\frac{5}{3}\dot{R}t. \tag{7.43}$$

In deriving these results, the thermal pressure of the gas ahead of the shock S_2 has been neglected compared with the 'ram' pressure, $\rho_0\dot{R}^2$. This is equivalent to saying that shock S_2 is a strong shock and these results are therefore valid as long as $\dot{R}$ is appreciably greater than the sound speed in the interstellar gas.

7.2.5 Efficiency of Conversion of Wind Kinetic Energy to Gas Kinetic Energy

We are now in a position to estimate the efficiency of conversion of mechanical wind energy into kinetic energy of surrounding gas. The kinetic energy, E_K, of the thin shell is given by

$$E_K=\frac{2}{3}\pi R^3\rho_0\dot{R}^2. \tag{7.44}$$

On using the expressions for R and $\dot{R}$ given by equations (7.41) and (7.42), this becomes

$$E_K = \frac{15}{77}\dot{E}_* t. \qquad (7.45)$$

The total mechanical energy output of the star up to time t is just $\dot{E}_* t$. Thus, a fraction 15/77 (i.e. about 20%) of the time-integrated mechanical energy output of the star is converted into kinetic energy of interstellar gas. This is much higher than the efficiency factor for converting radiant energy in the stellar UV into kinetic energy and compensates for the big difference (Section 7.2.1) in the stellar energy output rates in the two forms.

7.2.6 Does the Wind Impact on Ionized Gas?

It now remains to examine the assumption made in Section 7.2.2 that the interstellar gas around the star is ionized and not neutral. Let us suppose the star can produce a photoionized region before the wind starts to blow. We shall neglect any expansion of this region and assume that it has a uniform density n_0. When the wind blows, the interstellar gas is swept up into a thin shell of uniform density n_s. Since this gas cools until it becomes radiatively excited, the temperatures in the shell gas (*region* (*c*)) and surrounding interstellar gas (*region* (*d*)) will be about equal. Then, shock S_2 can be regarded as an isothermal shock and thus n_s is related to n_0 by the jump condition (equation (6.80))

$$n_s = \frac{\dot{R}^2}{c_i^2} n_0. \qquad (7.46)$$

where c_i is the sound velocity.

The thickness of the thin shell, say ΔR, can be found from mass conservation. The mass of gas in the thin shell must equal the mass of interstellar gas originally contained within a radius R. Hence,

$$4\pi R^2 n_s m_H \Delta R = \frac{4}{3}\pi R^3 n_0 m_H \qquad (7.47)$$

or

$$\Delta R = \frac{1}{3}\left(\frac{n_0}{n_s}\right) R. \qquad (7.48)$$

The recombination rate, $\dot{\mathcal{N}}_R$, in the whole shell therefore is given by

$$\dot{\mathcal{N}}_R = 4\pi R^2 n_s^2 \beta_2 \Delta R = \tfrac{4}{3}\pi R^3 n_0 n_s \beta_2 \quad s^{-1}. \qquad (7.49)$$

Note from equation (7.49) that the effect of compression has been to increase the recombination rate from the value it had when the gas was uniformly distributed with density n_0 m^{-3}. Now as long as $\dot{\mathcal{N}}_R$ is much less than the stellar

UV photon output rate S_*, the shell will not absorb a significant number of photons, and the gas outside shock S_2 will also be fully ionized. However, when $\mathcal{N}_R$ becomes about equal to S_*, no UV photons can escape into the surroundings and they become absorbed entirely in the shell. The approximate condition for this to happen is, therefore,

$$\frac{4}{3}\pi R^3 n_0\, n_s \beta_2 = S_*. \tag{7.50}$$

Using equations (7.41), (7.42), and (7.46), equation (7.50) can be used to derive a time t_R at which this occurs, given by

$$t_R \approx \frac{154}{60}\frac{c_i^2 m_H S_*}{\dot{E}_* n_0 \beta_2}\ \text{s}. \tag{7.51}$$

For an illustrative calculation, we take our previous standard values, namely $c_i = 10$ km s^{-1}, $S_* = 10^{49}$ s^{-1}, $\beta_2 = 2 \times 10^{-19}$ m^3 s^{-1}, $n_0 = 10^8$ m^{-3} together with $\dot{E}_* = 10^{29}$ J s^{-1} and obtain $t_R \approx 2 \times 10^{12}$ s $\approx$ 68 000 yr. The values of R and $\dot{R}$ obtained from equations (7.41) and (7.42) at this time and under these same conditions are respectively 2 pc and 20 km s^{-1}. We see, therefore, that our assumption – that the gas outside the shock is ionized – is reasonable until the outer shock velocity has fallen to about twice the sound speed in the ionized gas.

Although both this process and the thermal expansion process discussed previously in Section 7.1 impart kinetic energy to surrounding interstellar material, important differences between them should be noticed. The stellar wind gives kinetic energy to (mainly) ionized gas and the velocities are much higher than the sound speed there. Further, the gas is pushed into a shell-like configuration. On the other hand the expansion process deposits most of the kinetic energy in a shell of neutral gas which moves at a velocity much lower than the sound speed in the ionized gas. The ionized gas region is spherical but of roughly uniform density.

It is clear from observation that both processes occur in many galactic nebulae. However, because of complications caused by departures from spherical symmetry, non-uniformity of interstellar gas, and so on, our descriptions are good only in a qualitative way.

Figure 7.5 shows an example of a wind-blown bubble.

7.3 STELLAR WINDS FROM LESS MASSIVE STARS

As we have seen, the effects of massive stars (O and B types) on their interstellar environments are dramatic. However, these massive stars are very rare in number. Most stars have masses of about a solar mass or less, and there are very important differences between the way these less powerful stars interact with their surroundings and the way very massive stars interact. The three factors which cause these differences are:

FIGURE 7.5 An image of the Bubble Nebula. The bubble is about 2 kpc from Earth, and is about 2 pc in diameter. It is filled with very hot gas blown out by the wind from a single massive star. (Credit: NASA, ESA, and the Hubble Heritage Team (STScI/AURA).)

a These lower mass stars have negligible UV radiation fields.
b Their wind velocities are very much lower than those of O stars, typically $V_* \simeq 300$ km s^{-1} rather than some thousands of km s^{-1}.
c The wind mass loss rates are appreciably lower than those of O stars.

The flow pattern set up when the wind interacts with its surroundings is that sketched in Figure 7.3, but we must now note some very important changes in the physics of the four flow regions from the discussion in Section 7.2.3. *Region* (*a*) is occupied by unshocked stellar wind gas moving with velocity $V_* (\approx 300$ km s$^{-1})$. *Region* (*b*) contains shocked gas but there is now a crucial change. Because the wind velocity has been reduced by a factor of 10 or so relative to the O stars, the shocked wind temperature $\left(\propto V_*^2\right)$ is reduced by a factor of 100, that is, $T_s \approx 4 \times 10^5$ K. If we now refer to the cooling curve for shocked gas of cosmic abundances (Figure 6.6), we see that the shocked wind temperature is now near the peak cooling rate of this curve. Consequently, the cooling function $\Lambda(T)$ takes a much higher value than it does in the shocked wind gas of the very massive stars. The cooling time in shocked gas depends also on the pre-shock density (equation 6.69). In Section 7.2.3, we saw that the gas number density at distance r from a star with mass loss rate $\dot{M}_*$ at velocity V_* varies as $\dot{M}_*\left(V_* r^2\right)^{-1}$. As an example, we take the mass loss rate in a low mass star to be a tenth of that of a high mass star and take a similar ratio for the wind velocities. Thus, the pre-shock densities in the two winds are comparable for wind shocks occurring at a given distance from either star. The ratio of the cooling time in the shocked wind of the low mass star to that in the shocked wind of the high mass star is therefore 1:1000 since the cooling time varies as the cube of the wind velocity (Section 6.4.2).

We now have a very different situation to that discussed in Section 7.2.3. The radiative cooling in the shocked wind is now so efficient that to a good approximation, the shocked wind loses all its thermal energy immediately it passes through the shock. Consequently, the adiabatic expansion of shocked wind gas does not drive the shell of swept-up gas (*region* (c)). Rather this expansion is driven by the momentum in the wind. This is often called a *momentum driven flow*. In this case, the momentum in the flow at any time t is just equal to the total momentum in the stellar wind which has been ejected up to time t. This is in distinction to the energy driven flow produced by the wind of a massive star. In this latter case, the pressure of the shocked wind does work on its surroundings and can thereby generate momentum. *Region* (c) is a shell of shocked molecule-bearing gas driven by the wind momentum. It is swept up from molecular cloud material, *region* (d). It is thin because the gas cools well. We note in passing that the heating of molecular gas which passes through a shock can destroy molecules such as H_2 and CO since they are relatively weakly bound (binding energies of a few eV). However, it turns out that, provided a strong enough interstellar magnetic field is present, molecules can survive passage through shocks of velocity up to about 45 km s^{-1}.

7.3.1 Dynamics of Momentum Driven Flows

We examine the dynamics of momentum driven bubbles using the model sketched in Figure 7.6. The thin shell of radius R_s is bounded by the two shocks, S_1 and S_2. We neglect the mass of the stellar wind, which is always small compared with the mass of the swept-up gas. The rate of change of momentum of the swept-up material is now equal to the rate at which momentum is injected by the wind. The momentum injection rate, $\dot{\mu}_*$, of a wind of mass loss rate $\dot{M}_*$and velocity V_* is just

$$\dot{\mu}_* = \dot{M}_* V_*. \tag{7.52}$$

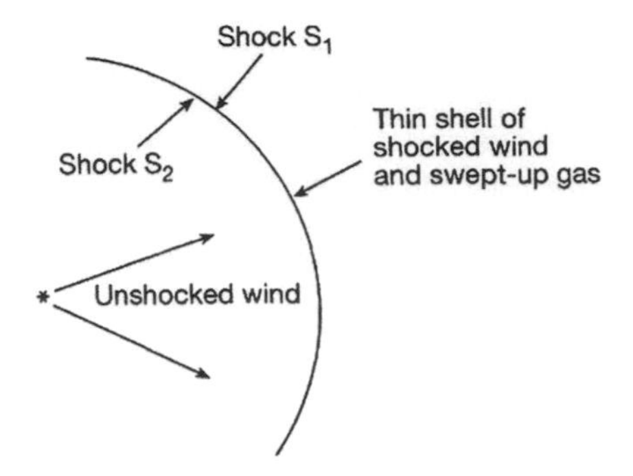

FIGURE 7.6 A momentum-driven wind-blown bubble.

The momentum conservation equation is therefore

$$\frac{\mathrm{d}}{\mathrm{d}t}\left(\frac{4}{3}\pi\rho_0 R_s^3 \dot{R}_s\right) = \dot{\mu}_*. \tag{7.53}$$

This can be integrated to give

$$\frac{4}{3}\pi\rho_0 R_s^3 \dot{R}_s = \dot{\mu}_* t \tag{7.54}$$

where we have put the integration constant to zero. Equation (7.54) also can be integrated to give

$$R_s = \left(\frac{3}{2\pi}\frac{\dot{\mu}_*}{\rho_0}\right)^{1/4} t^{1/2} \tag{7.55}$$

which satisfies the requirement that as $t \to 0$, $R_s \to 0$. Hence, the shell velocity is given by

$$\dot{R}_s = \frac{1}{2}\left(\frac{3}{2\pi}\frac{\dot{\mu}_*}{\rho_0}\right)^{1/4} t^{-1/2} \tag{7.56}$$

and obviously

$$R = 2\dot{R}_s t. \tag{7.57}$$

As characteristic parameters, we take $\dot{M}_* = 10^{-7} M_\odot \mathrm{yr}^{-1}$, $V_* = 300$ km s^{-1}, giving $\dot{\mu}_* = 2 \times 10^{21}$ kg m s^{-2}. Since molecular clouds are relatively dense, we take $n_0 = 10^9$ m^{-3} and put t_3 as the time in units of 10^3 yr. Equations (7.56) and (7.57) then give

$$R_s = 0.07 t_3^{1/2} \quad \mathrm{pc} \tag{7.58}$$

$$\dot{R}_s = 11 t_3^{-1/2} \quad \text{km s}^{-1}. \tag{7.59}$$

At times $t_3 \approx 1$, we obtain dimensions and velocities of flows that are quite consistent with what is actually observed.

7.3.2 Energy Conversion in a Momentum Driven Flow

Since, by definition, the wind energy is radiated away in a momentum driven flow, the conversion efficiency, f, of wind mechanical luminosity into shell kinetic energy must be low. This efficiency is given by

$$f = \frac{4}{3}\pi\rho_0 R_s^3 \dot{R}_s^2 / \dot{M}_* V_*^2 t. \tag{7.60}$$

From equations (7.55), (7.56), and (7.58), therefore,

$$f = \dot{R}_s / V_*. \tag{7.61}$$

So if, for example, $V_* = 300 \text{kms}^{-1}$ and $\dot{R} = 10$ km s^{-1}, the conversion efficiency is about 3%. This should be compared to the 20% or so conversion efficiency for energy driven winds. In principle, we can distinguish between energy and momentum wind driven interactions if we can estimate the energy conversion efficiency.

7.4 SUPERNOVA EXPLOSIONS AND SUPERNOVA REMNANTS

7.4.1 Introduction

A star whose initial main sequence mass is greater than about eight times the mass of the Sun gives rise, at the end of its life, to one of the most spectacular of astronomical events, the supernova. This is characterized by a sudden very large increase in the stellar brightness followed by a relatively longer decay. The brightness increase is so great that a supernova in a distant galaxy can clearly be distinguished from the normally individually unresolvable stellar background. This event involves the explosive ejection of about half the star's mass into its surroundings with a velocity of about one hundredth of the speed of light. The total kinetic energy of the ejected material is typically 10^{43}–10^{44} J. The sudden injection of this vast amount of energy at what is effectively a single point in the interstellar medium has dramatic consequences.

7.4.2 A Simple Model of the Supernova Explosion

A very simple but useful way of investigating the effects of this explosion on the interstellar medium is to regard it as the instantaneous release of a large amount of energy, E_* ($\approx 10^{43} - 10^{44}$ J) at a point in interstellar gas of uniform density n_0 m^{-3}. Obviously the energy deposition will heat the gas near the explosion site to a very high temperature and pressure and it will expand. Regardless of the thermal state (neutral or ionized) of the interstellar gas, the expansion velocity is highly supersonic. A shock wave is immediately set up which moves into the surroundings and sweeps up gas. This shock is driven by the pressure (i.e. the thermal energy) of a hot expanding gas bubble.

We will investigate the effects of the expansion (in the initial stages) using the model sketched in Figure 7.7. The shock wave S has a radius R and velocity $\dot{R}$. It is driven by the pressure in the hot bubble of gas interior to the shock. The initial shock velocity, $\dot{R}$, is very high but drops as the bubble expands. Unlike the case of the stellar wind interaction, it is important to consider the phase when radiative cooling in the shocked interstellar gas (which now fills the bubble) is unimportant. This occurs because the gas density around the supernova site is generally much lower than around a star possessing a wind. The characteristic interstellar density here is about 10^6 atom m^{-3} as opposed to perhaps 10^8 atom m^{-3}, characteristic in the wind interaction. This results in a longer cooling time. Possible reasons for the lower interstellar density are connected with the fact that the supernova event occurs after the star has been in existence for a considerable period of time. It might have moved away from the dense gas which was associated with the region of star formation and which may have become an HII region. Alternatively, if the star was surrounded by an HII region, the expansion process discussed previously would have reduced the gas density at the explosion site. There are important structural differences between the bubble of hot gas here and in the wind case. Previously we found that the hot shocked wind gas was trapped interior to a shell of swept-up material which had an overall expansion velocity much less than the sound speed in the hot gas. This hot gas therefore had uniform pressure. In the present case, the shock wave can be expanding at a speed greater than the sound speed in the gas immediately behind the shock. There can thus be a pressure gradient inside the bubble. However, the central regions, which are shocked to a very high

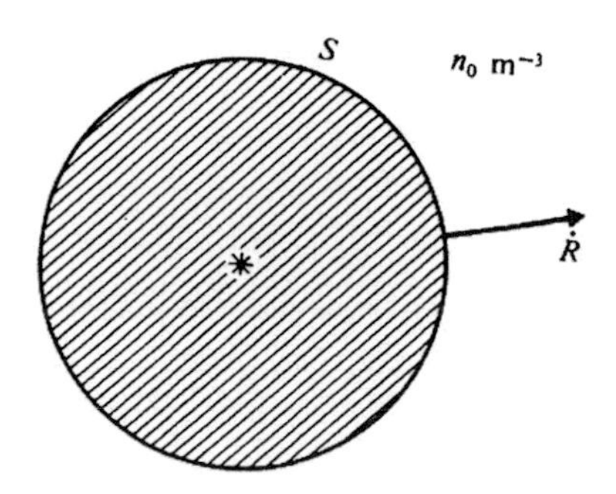

FIGURE 7.7 A schematic model of a supernova in interstellar gas.

temperature (and have a very high sound speed) in the initial expansion stages, are more nearly isobaric. We shall not attempt to take this structure into account in our treatment.

7.4.3 Radius and Expansion Velocity of the Supernova Bubble: The Energy-Conserving Phase

On the assumption of negligible radiative energy losses, conservation of energy demands that the total energy (kinetic plus thermal) of the gas behind the shock is equal to E_*, the explosion energy.

In Section 6.3.5, we saw that the kinetic (e_K) and thermal (e_T) energies per unit mass behind a strong adiabatic shock moving into a monatomic gas are equal and are given by

$$e_T = e_K = \frac{9}{32}\dot{R}^2. \tag{7.62}$$

In equation (7.62), the shock velocity is put equal to the time derivative of the radius since the gas ahead of the shock is taken to be at rest. As an approximation, we shall assume that the gas everywhere in the bubble has the specific energies given by equation (7.62). The total energy, E_T, of the gas in the bubble of density ρ_0 is given by

$$E_T = \frac{4}{3}\pi R^3 \rho_0 (e_T + e_K) = \frac{3}{4}\pi n_0 m_H R^3 \dot{R}^2. \tag{7.63}$$

Since

$$E_T = E_* \ , \tag{7.64}$$

equations (7.63) and (7.64) give as the equation of motion of the boundary of the bubble

$$R^3 \dot{R}^2 = \frac{4}{3\pi}\frac{E_*}{\rho_0}. \tag{7.65}$$

Since the bubble starts off at a very small radius, we take $R \to 0$ as $t \to 0$, as the boundary condition on equation (7.65). The solution of equation (7.65) is therefore

$$R = \left(\frac{25}{3\pi}\right)^{1/5} \left(\frac{E_*}{\rho_0}\right)^{1/5} t^{2/5} \tag{7.66}$$

so that

$$\dot{R} = \frac{2}{5}\left(\frac{25}{3\pi}\right)^{1/5} \left(\frac{E_*}{\rho_0}\right)^{1/5} t^{-3/5}. \tag{7.67}$$

Note that the power dependence of the radius and velocity on time differ from the wind case (equations (7.41) and (7.42)).

7.4.4 Radius and Expansion Velocity of the Thin Shell: The Momentum-Conserving Phase

As is evident from equation (7.67), the shock velocity decreases with time. Hence, the immediate post-shock temperature (which is proportional to the square of the shock velocity) also decreases with time. We noted in Section 6.4.1 that the radiative cooling rate in a gas may actually increase as its temperature decreases, because as the temperature falls there are more ions which are able to cool the gas via collisional excitation of lines. Radiative cooling of the gas immediately behind the shock therefore increases in importance as time goes on. Eventually, the rate of decrease of the gas temperature immediately behind the shock by radiative cooling can be greater than the rate at which it decreases by expansion. This gas tries to lose pressure but is pushed up against the shock by the very hot interior gas (which is so hot and tenuous that it does not cool appreciably over the timescales in which we are interested). This compression increases the gas density, which results in a further increase in the cooling rate. The cooling can be quite catastrophic and results in the formation of a thin shell of cool material immediately behind the shock. The situation is somewhat reminiscent of the wind case (Section 7.2.3), although there are again certain important differences. For example, in the supernova case, the thin shell contains only a fraction (though an appreciable one) of the interstellar gas which lay originally interior to the shock. Further, the push on the shell is produced by the pressure of hot interstellar gas heated in the very early stages. There is, of course, no source of energy addition to this hot gas. Finally, the decrease in the temperature of the gas in this thin shell is not restricted by the presence of a stellar UV radiation field. In principle, the gas could cool to a point at which it recombines to form neutral hydrogen, although this is complicated by the fact that energetic radiation can be produced both behind the shock wave and in the very hot interior gas.

We will make a very simple analysis of the motion of the shell during this stage by neglecting the push of the hot interior gas on the shell and by assuming that all the material interior to the shock resides in the shell. The thin shell is assumed to move outwards, sweeping up matter in such a way that the momentum of the shell is conserved. (This is often referred to as a 'snowplough' model.)

Since the gas cools so well the shell will be thin and so a single radius R and velocity $\dot{R}$ define its position and dynamics. Momentum conservation demands that

$$\frac{4}{3}\pi R^3 \rho_0 \dot{R} = \mathcal{M}_0 = \text{constant}. \tag{7.68}$$

Suppose that the thin shell is formed instantaneously at a time t_0 when $R = R_0$ and $\dot{R} = \dot{R}_0$. Then,

$$\mathcal{M}_0 = \frac{4}{3}\pi R_0^3 \dot{R}_0 \rho_0. \tag{7.69}$$

On integration, equation (7.58) therefore gives

$$R = R_0\left(1 + 4\frac{\dot{R}_0}{R_0}(t - t_0)\right)^{1/4} \tag{7.70}$$

and

$$\dot{R} = \dot{R}_0\left(1 + 4\frac{\dot{R}_0}{R_0}(t - t_0)\right)^{-3/4} \tag{7.71}$$

For large enough times ($t \gg R_0/\dot{R}_0$), $R \propto t^{1/4}$ and $\dot{R} \propto t^{-3/4}$, characteristic of momentum-conserving solutions. Comparison with the power dependence for the energy-conserving phase (equations (7.66) and (7.67)) shows, as would be expected, that the radius increases more slowly and the velocity decreases more rapidly with time in the momentum-conserving stage. (The inclusion of the interior pressure produces a time exponent somewhere between $\frac{1}{4}$ and $\frac{2}{5}$ in R.)

An estimate of the parameters R_0, $\dot{R}_0$, and t_0 demands a more detailed investigation of the cooling of gas heated to high temperatures by passage through a shock wave. This is beyond the scope of this book. However, we will state without further justification that it occurs when $\dot{R}_0$ typically is a few hundred km s^{-1}.

During the adiabatic stage of evolution ($t < t_0$), the gas temperature is high ($T \gtrsim 10^6$ K). Most of the radiation comes out in the far UV or as soft X-rays. X-ray maps of young supernova remnants suggest a broad agreement between the theory and observation. The situation changes once $t \gtrsim t_0$. Because the cooling is so catastrophic, gas temperatures $\lesssim 10^5$ K are reached in the thin shell. The gas then becomes a copious emitter of collisionally excited lines (such as those from O^{++}). By then the remnant has a diameter of several tens of parsecs and is optically visible as a very thin shell. Figure 1.3(b) shows a spectacular example of the radiating gas behind a supernova blast wave.

7.4.5 Numerical Estimates

In order to make some quantitative estimates regarding the remnant dynamics, we will adopt as representative values $E_* = 10^{44}$ J and $n_0 = 10^6$ m^{-3}. Equations (7.66) and (7.67) then give $R \approx 3.6 \times 10^{-4} t^{2/5}$ pc and $\dot{R} \approx 4.4 \times 10^9 t^{-3/5}$ km s^{-1} (where t is measured in seconds). As a typical value for the velocity at the transition point for catastrophic cooling, we take $\dot{R}_0 = 250$ km s^{-1}. The transition time, t_0, and radius, R_0, can be estimated respectively as 39 000 yr and 24 pc. The amount of interstellar matter swept up at this time is $1400 M_\odot$, which is several orders of magnitude greater than the ejected mass. Note that, with our adopted value of n_0, about $4 M_\odot$ of interstellar material has been swept up by the time the remnant has reached a radius of about 3.4 pc, that is, very early on in the evolution. This is some justification for neglecting the mass of the ejecta ($\approx 4 M_\odot$) and assuming all the energy is transferred to interstellar gas.

7.4.6 Efficiency of Energy Conversion

We can now use the previous results to estimate the efficiency with which the explosion energy has been converted into gas kinetic energy. We derived our results for the adiabatic stage using the simplifying assumption that the explosion energy was shared equally between the kinetic and thermal energy of swept-up gas. More sophisticated treatments show that in fact about 30% of the total energy appears as kinetic energy and the remainder as thermal energy. When this is taken into account, the numerical coefficients of equations (7.66) and (7.67) change slightly, but the changes are unimportant for our estimates.

In the snowplough phase, the fraction of the explosion energy appearing as kinetic energy is

$$g \approx \frac{2\pi}{3E_*}\rho_0 R^3 \dot{R}^2 \tag{7.72}$$

where we have assumed that the mass in the cool shell is equal to the swept-up mass. The use of equations (7.70) and (7.71) gives this fraction at large enough times to be

$$g \approx \frac{4^{1/4}\pi}{6E_*}\rho_0 R_0^{15/4} \dot{R}_0^{5/4} t^{-3/4} \, . \tag{7.73}$$

If we again use the values of R_0, $\dot{R}_0$, ρ_0, and t_0 of Section 7.3.5, then $g \approx 0.6(t_0/t)^{3/4}$. For example, if $\dot{R} = 10$ km s^{-1}, then $t/t_0 \approx 44$ and the resulting conversion efficiency is just under 4%, that is, rather low.

Note, however, that not all of the rest of the explosion energy has been lost to the interstellar medium. The very hot gas produced in the early stages of the expansion cools very slowly (because of its high temperature and low density). Its expansion can result in further transference of energy (ultimately derived from the explosion energy) to interstellar gas.

7.4.7 The Very Early Stages of Supernova Remnant Evolution

As noted in Section 7.4.5, the mass of swept-up material exceeds that of the ejecta at a radius (or time) much less than that at which strong cooling sets in. On simple momentum considerations, we would expect the ejecta velocity at this time to be about equal to its initial velocity. Hence, the characteristic time at which this occurs is $t_e \approx R_e/V_e$, where R_e is the radius at which the swept-up and ejected masses are equal and V_e is the initial velocity of the ejecta. Using the value $R_e \approx 3.4$ pc (Section 7.4.5) and typical $V_e = 5000$ km s^{-1}, then $t_e \approx 600$ yr. Although $t_e \ll t_0$, the evolution of the remnant up to time t_e is of great interest, since such very young remnants are observed.

We will therefore briefly describe this initial interaction in a qualitative way. The ejected stellar material acts as a supersonic piston, which drives a strong shock into the surrounding material. (This is, of course, the shock whose motion has been discussed at length previously.) The outflowing ejecta are slowed down by this process. This retardation is caused by a second shock which moves inwards through the ejected material. We have, therefore, a two-shock flow pattern analogous to that shown in Figure 7.3. In terms of the regions in that figure, *region* (*a*) is unshocked and *region* (*b*) shocked ejecta. Detailed investigation shows that the region of flow between the two shocks is highly unstable and leads to radial 'tongues' of shocked ejecta penetrating the shocked interstellar gas. In doing so, the interstellar magnetic field which is present in the swept-up material is bent back more or less into the radial direction. Direct evidence of this magnetic field deformation comes from radio-frequency observations of the synchrotron radiation produced by relativistic electrons moving under the influence of these magnetic fields. The origin of the relativistic electrons is related to the complex motions of gas and magnetic fields in this unstable flow. The instability eventually dies out and the subsequent evolution follows the outline of the previous sections.

Finally, it is perhaps worth noting that the instability also occurs in the two-shock stellar wind flow pattern. However, it dies out so early in the evolution that it is not important in the context of the discussion presented. Figure 7.8 shows a portion of the supernova remnant called the Veil Nebula in the constellation of Cygnus.

FIGURE 7.8 An image of part of the Veil Nebula, a supernova remnant resulting from the explosion of a 20 solar mass star some 8000 years ago. The explosion created a bubble of hot gas expanding into a previously wind-swept cavity in cool interstellar gas. The image shows a small portion of the interaction zone; hydrogen emission is shown in red, sulfur in green, and oxygen in blue. (Credit: NASA/ESA/Hubble Heritage Team.)

7.5 CONSEQUENCES OF SUPERNOVAE FOR THE INTERSTELLAR MEDIUM OF GALAXIES

7.5.1 THE PRODUCTION OF CORONAL GAS

This gas component of the interstellar medium is extremely hot and is named by analogy with the hot gas in the solar corona. The most obvious evidence of its existence (and a pointer to its origin) comes from observations of X-rays from known sites of supernovae. Soft X-rays (energies roughly in the range 0.1–2 keV) are detected, implying gas temperatures in excess of 10^6 K. An indication of even more extensive lower-temperature (approximately 5×10^5 K) gas has come from satellite observations of the ultraviolet spectra of hot stars. The spectra of such stars often show absorption lines produced by the O VI ion (i.e. oxygen with five electrons detached) so that the hot gas must lie along the line of sight. The density of this coronal gas is low, less than 10^4 particle m^{-3} being reasonably representative.

The excitation and ionization of this gas must be due to collisional processes since photoionization produces temperatures only of about 10^4 K. The most likely source of deposition of energy into the gas is via a shock wave, and a supernova blast wave is the obvious candidate. We noted in Section 7.3.6 that the very hot gas produced in the initial adiabatic phase of evolution of a supernova remnant cools very slowly. This is in fact the coronal gas. Because it is so hot, it has a very large scale height in the galactic gravitational field, much greater than that of neutral gas. It occupies about half the volume of interstellar space in the Milky Way galaxy. For reference, some properties of this and other components of the interstellar medium are listed in Table 2.1, Section 2.5.

7.5.2 THE EFFECTS OF GROUPS OF MASSIVE STARS ON THE INTERSTELLAR MEDIUM OF GALAXIES

So far we have discussed the effects of a wind from a single star and a single supernova explosion on the interstellar medium, and thus have considered them to act quite independently of each other. In fact, as will be described in Chapter 8, massive stars are often found in groups. The number of stars in a group can range from a few up to many million (in external galaxies). In any group, there is a spread in spectral type of the massive stars, that is, in their masses. Consequently there is a spread in the rate at which they evolve, and some stars can become supernovae while others are still blowing their winds. Thus, the joint action of winds and supernovae should, in principle, be considered. However, studies of the energy input from winds and supernovae in groups of stars show that, to a good approximation, once supernovae occur, they dominate the energy input. We will take this point of view here.

To a first approximation, the effects of injection of a supernova energy E_* every time step τ can be considered to be the equivalent of a wind of mechanical luminosity $\dot{E}_*\equiv E_*/\tau$. We can, therefore, use equations (7.41) and (7.42) to investigate the dynamics of the bubble driven by these repeated supernova explosions. To see the effects of this, we take $E_* = 10^{44}$ J and $\tau = 3 \times 10^5$ yr, giving an effective wind mechanical luminosity $\dot{E} = 10^{31}$ W. (This is roughly equivalent to the simultaneous wind input from 100 or so massive O stars – Section 7.2.1.) Inserting these values into equations (7.41) and (7.42) and defining t_7 as the time in units of 10^7 years gives a bubble radius and velocity respectively of

$$R \simeq 4.3 \times 10^3 t_7^{3/5}/n_0^{1/5} \quad \text{pc} \tag{7.74}$$

$$\dot{R} \simeq 260/t_7^{2/5} n_0^{1/5} \quad \text{kms}^{-1} \,. \tag{7.75}$$

Taking our usual value for the interstellar density in a typical galaxy of $n_0 = 10^6$ m^{-3}, then after a time of 10^7 years, the bubble has a radius $R \simeq 270$ pc and velocity $\dot{R} \simeq 16$ km s^{-1}. Thus, groups of stars can blow very large bubbles ('super-bubbles') in galaxies. Such bubbles can be seen in photographs of external galaxies. Our neighbouring galaxy, the Large Magellanic Cloud, contains a spectacular example of a super-bubble (see Figure 7.9).

FIGURE 7.9 Image of the super-bubble Henize 70 (or N70) in the Large Magellanic Cloud. This false colour image shows the super-bubble created by supernovae winds. The bubble is about 100 pc in diameter and filled with very low density, very high temperature gas. (Credit: ESO www.eso.org/public/images/eso9948d/.)

We must be careful in the application of equations (7.74) and (7.75), which assume a uniform interstellar gas density. In fact, the density of interstellar gas decreases with increasing distance from the galactic plane, so the interstellar gas distribution is stratified, exactly as in a hydrostatic atmosphere. In a typical spiral galaxy, the scale height of this gas is about 100 pc, that is, very significantly less than the radius of the bubble calculated after a time of about 10^7 yr.

The effect of this density variation has important consequences for the evolution of the bubbles. Once a bubble has expanded to a distance about equal to the scale height of the density distribution of the interstellar gas, the bubble dynamics start to deviate strongly from those of bubbles expanding into gas of uniform density. The reason is clear from an inspection of equations (7.74) and (7.75). First, consider that section of a bubble which is expanding along the plane of the galaxy. There the density is constant and equations (7.74) and (7.75) are applicable. Now consider that section expanding perpendicular to the plane of the galaxy. As the shell moves away from the plane, the density decreases and clearly, for equations (7.74) and (7.75), at any particular instant of time, this section will have a greater radius and higher velocity than that section moving along the plane of the galaxy. The shell elongates and distorts as sketched in Figure 7.10. As the evolution continues, the bubble can actually burst as sketched, and the gas that was driving the bubble can rise well above the galaxy and produce a hot 'halo'.

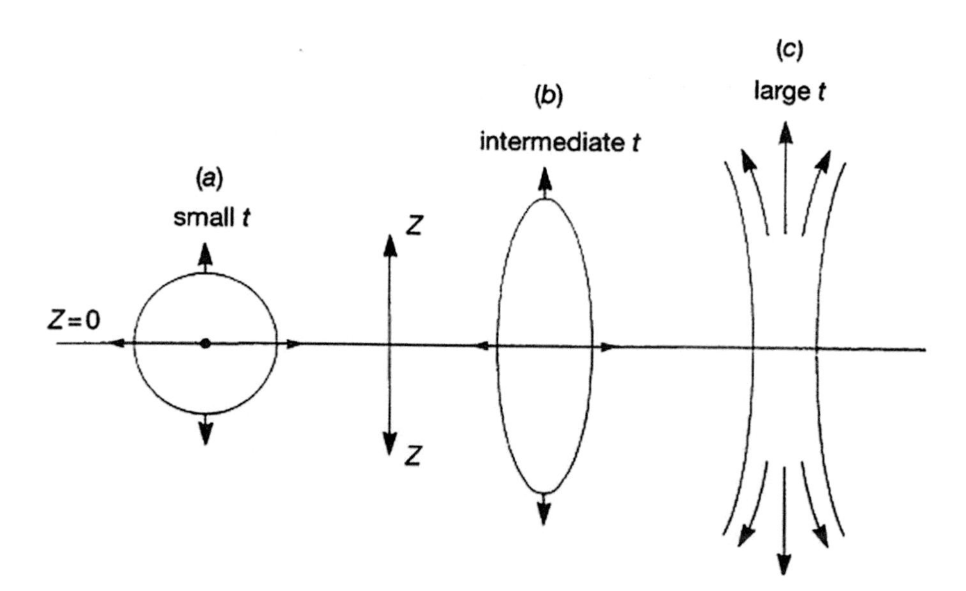

FIGURE 7.10 The distortion of a wind-blown bubble. The interstellar gas density decreases in the z direction.

PROBLEMS

1. The Strömgren spheres around two stars emitting 10^{47} and 10^{49} UV photon s^{-1}, respectively, are observed to be in pressure equilibrium with their surroundings. What is the ratio of the interstellar gas densities around the two stars if the Strömgren spheres have identical radii?
2. A nebula of radius 1 pc and density 10^8 m^{-3} has a star cluster containing $100M_\odot$ material at its centre. Would gravity affect the expansion of the nebula?
3. A star emits a stellar wind with an energy output rate of 10^{29} J s^{-1} into interstellar gas of density 10^7 m^{-3}. Calculate the gas density and the thickness of the region of swept-up interstellar gas after 30 000 yr. (Assume that the outer shock is isothermal and the gas temperature on either side is 10^4 K.)
4. A spherical bubble of hot gas headed by a strong shock expands into gas whose density is inversely proportional to the square of the distance from the bubble centre. An energy source ensures that the total energy of the hot gas remains constant. Find the dependence on time of the shock radius and velocity. (Assume that all the energy of the hot gas is in the form of thermal energy.)
5. A supernova explosion liberates 10^{44} J in interstellar gas of density 10^6 m^{-3}. Estimate what fraction of the initial energy has been radiated away by free–free radiation after 10 000 yr. (Assume that the gas density and temperature in the hot bubble are uniform and equal to their post-shock values. Energy is lost by free–free emission at a rate $L_{ff} = 1.4 \times 10^{-40} T_e^{1/2} N_e^2$ J $m^{-3}s^{-1}$, where T_e and N_e are respectively the electron temperature and density.)
6. A stellar wind drives a bubble of gas, which is observed to expand at a constant velocity. Use dimensional arguments to find how the gas density around the star varies with the radial distance from the star.
7. A supernova of energy 10^{44} J explodes in interstellar matter of density $n_0 = 10^6$ m^{-3} and pressure $P_I = 3 \times 10^{-14}$ N m^{-2}. Estimate how much gas can be disturbed by this supernova.

8 Star Formation and Star-Forming Regions

8.1 INTRODUCTION

The average density in a star is in excess of 10^3 kg m^{-3}, and this is enormously greater than any densities encountered in the interstellar gas. Evidently, great compression must occur, and the only force capable of producing this in a mass of gas is the self-gravity of the gas. However, gravity must overcome a variety of disruptive forces. For example, gas and magnetic pressure, turbulence and rotation, all act against compression. Indeed, so many are the effects opposing star formation that the difficulties seem almost insuperable. However, it is obvious from the structure of the Galaxy that Nature has no such difficulties!

The existence of stars which have lifetimes much less than the age of the Galaxy implies that star formation must be an ongoing process. This is further suggested by heavy element abundances, which clearly show that many cycles of nuclear burning must have occurred. We will consider only the simplest possible situations involving star formation and look at necessary – but not sufficient – conditions which must be satisfied for star formation to take place.

8.1.1 THE EQUILIBRIUM OF A SINGLE CLOUD

We first consider a single isolated spherically symmetric cloud in equilibrium under three forces, namely internal pressure, self-gravity, and surface pressure exerted by an external medium. To establish the necessary equations, consider a spherical shell in the cloud of thickness dr at radius r (Figure 8.1). The shell has mass

$$\mathrm{d}M(r) = 4\pi r^2 \rho(r)\mathrm{d}r \tag{8.1}$$

where $\rho(r)$ is the density of the gas at radius r. In equilibrium, the inwards gravitational force on the shell due to the mass $M(r)$ interior to it must be balanced by a pressure gradient. Thus, a pressure differential d$P(r)$ must exist across the shell. Obviously, the pressure must decrease outwards to produce an outwards force. For equilibrium, therefore,

$$4\pi r^2 \mathrm{d}P(r) = -GM(r)\mathrm{d}M(r)/r^2. \tag{8.2}$$

We can write this equation as

$$3V(r)\mathrm{d}P(r) = -GM(r)\mathrm{d}M(r)/r \tag{8.3}$$

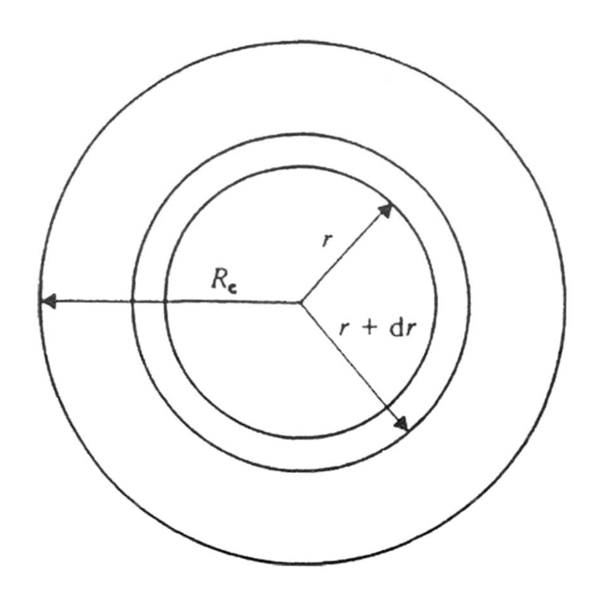

FIGURE 8.1 Geometry of a spherical shell.

where $V(r) \equiv \frac{4}{3}\pi r^3$ is the interior volume at radius r. Let us now integrate equation (8.3) from the centre of the cloud to its edge where it has radius R_c and pressure equal to the external pressure P_s. Thus

$$3 \int_{P_{c0}}^{P_s} V(r)\mathrm{d}P(r) = -\int_0^{M_c} \frac{GM(r)\mathrm{d}M}{r}. \tag{8.4}$$

In equation (8.4), M_c is the total mass of the cloud and P_{c0} is the central pressure in the cloud. The left-hand side of equation (8.4) can be integrated by parts, giving

$$3 \int_{P_{c0}}^{P_s} V(r)\mathrm{d}P(r) = 3V(r)P(r)|_{\text{centre}}^{\text{edge}} - 3\int_0^{V_c} P\mathrm{d}V = 3V_cP_s - 3\int_0^{V_c} P\mathrm{d}V. \tag{8.5}$$

Here $V_c \equiv \frac{4}{3}\pi R_c^3$ is the volume of the cloud.

Since the internal energy, ε_i, per unit volume of monatomic gas is given by

$$\varepsilon_i = \frac{3}{2}P \tag{8.6}$$

we can write

$$\int_0^{V_c} P\mathrm{d}V = \frac{2}{3}\int_0^{V_c} \varepsilon_i \mathrm{d}V = \frac{2}{3}\mathcal{T} \tag{8.7}$$

where $\mathcal{T}$ is the thermal energy content of the cloud.

The right-hand side of equation (8.4) is just the gravitational self-energy, Ω, of the cloud. Equation (8.4) can be written as

$$3V_c P_s = 2\mathcal{T} + \Omega. \tag{8.8}$$

Other forces (e.g. magnetic fields) could be added to this formulation.

8.1.2 The Collapse of an Isolated Gas Cloud and Spontaneous Star Formation

We will initially consider a spherical gas cloud, on which the only forces acting are those due to its self-gravity and internal pressure. At first„ we will ignore the surface pressure exerted by any surrounding gas, since the conclusions are not drastically changed by its inclusion, but will discuss its significance later (Section 8.1.3). We have already derived a criterion which must hold if the cloud is to be in equilibrium. Applied to these circumstances, it takes the form of equation (8.8) with the surface pressure term removed, that is,

$$2\mathcal{T} + \Omega = 0. \tag{8.9}$$

Suppose that we have a situation where $2\mathcal{T} > -\Omega$, that is, $2\mathcal{T} + \Omega > 0$. Then the pressure term dominates and intuitively we would expect the cloud to expand. Conversely, if $2\mathcal{T} < -\Omega$, that is, $2\mathcal{T} + \Omega < 0$, we would expect the cloud to be contracting. These intuitive expectations can be confirmed by analysis and we can write the following schematic equation to describe the dynamical state of an isolated cloud:

$$\begin{aligned} 2\mathcal{T} + \Omega &< 0 \quad \text{Contraction} \\ &= 0 \quad \text{Equilibrium} \\ &> 0 \quad \text{Expansion} \end{aligned} \tag{8.10}$$

Equation (8.3) shows that the pressure and density in the cloud will vary with radius. This variation can be found by integrating this equation subject to assumptions connecting the cloud pressure and density. In order to make simple estimates, we will assume that the cloud has a uniform density ρ_c and pressure P_c. For a uniform cloud then

$$2\mathcal{T} = 3P_c V_c \tag{8.11}$$

and

$$\Omega = -\int_0^{M_c} \frac{GM(r)\mathrm{d}M}{r} = -\frac{16\pi^2}{3}\rho_c^2 G \int_0^{R_c} r^4 \mathrm{d}r = -\frac{3}{5}\frac{GM_c^2}{R_c}. \quad (8.12)$$

Thus, in order to start the contraction which is necessary if a cloud is ultimately to form a star (or stars), we require

$$\frac{3}{5}\frac{GM_c^2}{R_c} \gtrsim 4\pi R_c^3 P_c. \quad (8.13)$$

Since

$$P_c = \frac{3M_c k T_c}{4\pi R_c^3 m_H \mu} \quad (8.14)$$

where T_c is the gas temperature (assumed constant), condition (8.13) becomes

$$\frac{GM_c}{5R_c} \gtrsim \frac{kT_c}{\mu m_H}. \quad (8.15)$$

Now, $kT_c/\mu m_H = c_c^2$, where c_c is the sound speed in the cloud (which will be assumed to behave isothermally). The sound travel time across the cloud is $t_s \approx R_c/c_c$. Hence equation (8.15) can be written as

$$t_s \gtrsim (15/4\pi G\rho_c)^{1/2}. \quad (8.16)$$

The term on the right-hand side of condition (8.16) has the following physical interpretation. Consider a spherical cloud which is allowed to collapse under self-gravity. If we ignore the effects of the internal pressure then this is said to be a free-fall collapse. The equation of motion of a thin shell situated at an initial distance r_0 from the centre (Figure 8.1) is just

$$\frac{\mathrm{d}^2 r}{\mathrm{d}t^2} = -\frac{4}{3}\frac{\pi G r_0^3 \rho_c}{r^2}. \quad (8.17)$$

In deriving this equation, we have assumed that the cloud initially has a uniform density ρ_c. We define the following quantities:

$$x = r/r_0, \quad \tau = t/t_{ff}, \quad t_{ff} = \sqrt{3\pi/32G\rho_c}. \quad (8.18)$$

Equation (8.17) then takes the form

$$\ddot{x} = -\frac{\pi^2}{8x^2} \quad (8.19)$$

where the dot indicates differentiation with respect to τ. We can put equation (8.19) into the form

$$\frac{\mathrm{d}}{\mathrm{d}x}\dot{x}^2 = -\frac{\pi^2}{4x^2} \tag{8.20}$$

and integrate to get

$$\dot{x}^2 = \frac{\pi^2}{4x} + a. \tag{8.21}$$

Here a is a constant of integration. If the cloud starts off from rest, then when $x = 1$ $\dot{x} = 0$and equation (8.21) becomes

$$\dot{x} = -\frac{\pi}{2}\left(\frac{1}{x} - 1\right)^{1/2}. \tag{8.22}$$

(Note that we take the negative root since $\dot{x}$ is directed inwards.) In order to integrate this equation, set $x = \cos^2\theta$. Then equation (8.22) becomes

$$\cos^2\theta\,\dot{\theta} = \frac{\pi}{4} \tag{8.23}$$

which can be immediately integrated to give

$$\frac{\theta}{2} + \frac{1}{4}\sin 2\theta = \frac{\pi\tau}{4} + b \tag{8.24}$$

where b is the constant of integration. When $\tau = 0$, $x = 1$ so that $\theta = 0$. Hence $b = 0$. Now the shell reaches the centre when $x = 0$, that is, when $\theta = \pi/2$. From equation (8.24) we thus obtain $\tau = 1$ when $x = 0$. Note that this time is the same for all r_0. Hence the free-fall collapse time is just

$$t_{\mathrm{ff}} = \left(\frac{3\pi}{32G\rho_{\mathrm{c}}}\right)^{1/2}. \tag{8.25}$$

We now see that the condition for collapse (equation (8.16)) can be written in the form

$$t_s \gtrsim \frac{2\sqrt{10}}{\pi}t_{\mathrm{ff}} \approx 2t_{\mathrm{ff}}. \tag{8.26}$$

Physically then the condition for collapse is that the free-fall time must be less than about the time taken for a sound wave to cross the cloud. We can write equation (8.26) in an alternative form by defining M_{crit} as the critical mass for collapse of a cloud of density ρ_{c} and internal sound speed c_{c}. This critical mass is often referred to as the Jeans mass. Then equation (8.26) gives the collapse condition as

$$M \gtrsim M_{\text{crit}} = \left(\frac{3\pi^5}{32}\right)^{1/2} c_c^3 G^{-3/2} \rho_c^{-1/2}. \tag{8.27}$$

Let us now calculate the critical mass under the density and temperature conditions typical of the diffuse neutral clouds discussed in Section 7.4.1. We find $M_{\text{crit}} \approx 10^4 M_\odot$. This is much greater than their observed masses. We conclude that star formation does not take place in such clouds and ample evidence confirms this conclusion.

If we apply these ideas to the gas in cool molecular clouds, we come to rather different conclusions. We will assume that the typical particle density is $5 \times 10^9\ \text{m}^{-3}$ and the gas temperature is 20 K. The critical mass then is $M_{\text{crit}} \approx 30 M_\odot$, which is derived assuming that the mean mass of a particle is $2m_{\text{H}}$. The critical radius corresponding to this mass is $R_{\text{crit}} \approx 0.3$ pc. Thus a given molecular cloud could contain many subunits which can collapse independently. Hence, if we assume that the type of collapse discussed is necessary for at least some chance of star formation, we would conclude that it is likely that stars do not form individually but in groups within a parent cloud of high mass. This expectation is generally borne out by observation.

However, once a given cloud starts to contract, the story does not end. Equation (8.27) shows that provided c_c remains constant (or decreases with increasing density), the critical mass for collapse decreases as the collapse proceeds (because the average density increases). We therefore expect that the initially collapsing cloud (which may be a subsection of a more massive cloud) becomes liable to break up into fragments of smaller mass, which could themselves collapse independently. These fragments could become liable to fragmentation later on by the process by which they were formed.

This process of break-up into a hierarchy of fragments cannot continue indefinitely. As a cloud collapses, part of its gravitational potential energy can be converted into heat by compression or by the generation of supersonic motions, which transform kinetic energy into heat via shock waves. If this energy input increases the sound speed, then we can see from equation (8.27) that M_{crit} would be increased and the successive process of fragmentation ultimately will be halted. Generally, this occurs when the heat generated can no longer be radiated away efficiently. This will occur when the cooling radiation is unable to escape because of increasing opacity. Detailed investigation of this process suggests that the lowest mass of fragment that can be formed by successive fragmentation is about 0.01 $M_\odot$.

8.1.3 Induced Star Formation

We next consider clouds which are initially in equilibrium, now including an external surface pressure, and which therefore satisfy equation (8.8). This equation can be written in the form

$$P_s = \frac{A}{R_c^3} - \frac{B}{R_c^4} \tag{8.28}$$

where we have used equations (8.11) and (8.12).

In equation (8.28), A ($\equiv 3kT_cM_c/4\pi\ \mu m_H$) and $B(\equiv 3GM_c^2/20\pi)$ are constants for a cloud of a given mass M_c and temperature T_c. A sketch of the variation of P_s with R_c is shown in Figure 8.2.

If we fix the external pressure at some value P_0, then the cloud can exist in either of the two equilibrium states, E and D (Figure 8.2). Suppose that the surface pressure is now raised to $P_0 + \Delta P_0$; we will briefly discuss later how this can happen. Figure 8.2 shows that, in principle, we move to new equilibrium states E' and D'. But we must be cautious about this. First, consider a cloud initially at D. Increasing the surface pressure physically must cause the cloud to contract, that is, R_c must decrease. The internal pressure in the cloud increases to balance the external pressure. Hence the cloud moves from $D \rightarrow D'$ with decreasing radius and increasing internal pressure. What happens if we start at E? If the external pressure is increased, Figure 8.2 shows that we will have to increase the cloud radius in order to achieve a new equilibrium at E'. This is physically not possible, since the increased external pressure must cause the cloud to contract. Hence the cloud cannot stay in equilibrium and it starts to collapse. In other words, cloud equilibria to the left of the maximum M ($R_c < 4B/3A$) in Figure 8.2 are unstable to collapse. Obviously, therefore, if we start with a stable configuration at D and increase the pressure sufficiently, we can cause clouds to move into an unstable regime. If this process leads to star formation, it is called 'induced' star formation, as opposed to 'spontaneous' star formation where a cloud collapses because it satisfies the contraction condition (8.10).

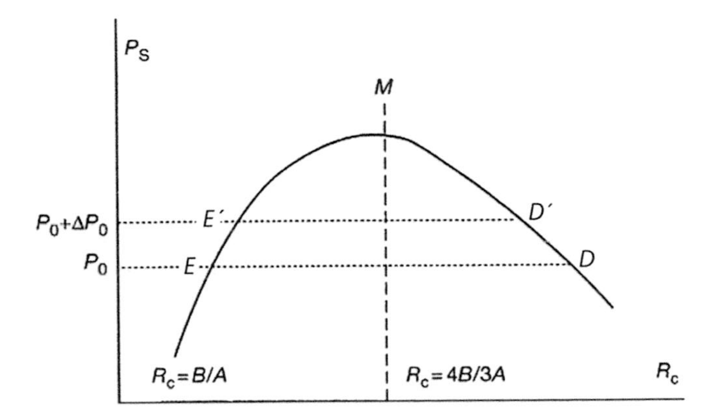

FIGURE 8.2 The pressure–radius relationship for an isothermal cloud with a surface pressure.

We have already met various dynamical processes that can increase the pressure of the interstellar medium, namely photoionization, supernovae explosions, and stellar wind activity. All these may play a role in inducing star-forming activity. The fact that the formation and dynamical effects of massive stars may induce further star formation gives rise to the concept of 'sequential star formation', in which successive generations of massive stars drive fast shocks into the neighbouring gas, triggering a new generation of star formation, so that the process repeats until the conversion of gas to stars is complete. The existence of chains of massive stars with the oldest at one end and the youngest at the other is strong evidence for these effects.

8.1.4 Other Factors Affecting Star Formation

The picture of star formation as presented is simplified almost to the point of unreality. We have, for example, neglected the effects of the internal structure of the cloud. There are even more important omissions. The collapsing gas cloud will contain a magnetic field [whose effects should be included in equation (8.8)]. Collapse increases the magnetic field strength and hence the magnetic pressure, and this will act, like thermal pressure, to oppose the collapse. The importance of magnetic fields depends critically on both the field geometry and how the field is coupled to the gas. Ions are tied to magnetic field lines, but neutral atoms and molecules are affected by magnetic fields only through collisions between ions and neutrals. Therefore, neutrals may drift past the ions in a process depending on the ion abundance called ambipolar diffusion. The ambipolar diffusion timescale, t_{amb}, depends on the fractional ionization, x_i, and is approximately $t_{amb} \sim 4 \times 10^5 (x_i/10^{-8})^{1/2}$ years.

We can gain some insight into the role of magnetic fields during gravitational collapse by comparing the ambipolar diffusion timescale with the free-fall timescale. If the ambipolar diffusion timescale is long compared to the free-fall timescale in years, $t_{ff} \sim 3 \times 10^{10}/n_H{}^{1/2}$ (where $n_H = [n(H) + 2n(H_2)]$ m^{-3} is the total hydrogen atom number density), then we conclude that magnetic fields may be affecting the gravitational collapse. For example, in diffuse clouds where the interstellar radiation field maintains x_i at about 10^{-4} in clouds with hydrogen number density $n_H \sim 10^8$ m^{-3}, then t_{amb} exceeds t_{ff}, so the magnetic field is locked in during the gravitational collapse and the magnetic pressure may be sufficient to arrest the collapse. In dark clouds of number density $\sim 10^{10}$ m^{-3} where cosmic rays maintain x_i at about 10^{-8}, t_{amb} and t_{ff} are comparable. For denser regions where dust grains and chemistry conspire to maintain even lower ionization fractions, the ambipolar diffusion timescale is short and the magnetic field may drift quickly out of the gas and play no further role in the collapse.

Other complicating effects include those of cloud rotation, since it is very unlikely that any collapsing cloud will have zero angular momentum. In the absence of a braking torque, the rotational velocity of a cloud will increase because angular momentum is conserved during the gravitational collapse. If the magnetic field is connected to surrounding gas, the necessary braking torque may be produced.

Star formation still remains one of the most fundamental of astrophysical problems. Fortunately, observations of star formation in regions that are inevitably deeply embedded in molecular clouds are now possible by means of far-infrared emission from dust grains and submillimetre emissions from molecules. We describe in Sections 8.2–8.4 the pictures of low-mass and high-mass star formation that are emerging from these observational studies.

8.2 OBSERVATIONAL SIGNATURES OF LOW-MASS STAR-FORMING ACTIVITY

We noted in Section 8.1.3 that gravitational collapse of a sufficiently massive sphere of gas may lead to the formation of a star, if other effects opposing collapse do not dominate. In particular, rotation of the collapsing cloud, the role of magnetic fields, turbulence, and heating and cooling effects were ignored in that calculation, yet it is possible that some or all of these may be important. In this section, we consider the observational evidence obtained from studies of low-mass star-forming regions to see if this information can help us to understand the role of these additional physical effects on the collapse. We then describe qualitatively how this evidence may be understood in terms of a more general picture of star formation.

8.2.1 MOLECULAR CLOUDS AND FILAMENTS

Cold molecular clouds are the interstellar reservoirs of mass from which new stars may form. These clouds can be probed by molecular line emissions and by the continuum emissions from dust. The dust emissions occur in the far infrared and reveal the remarkable result that interstellar gas in these cold clouds is arranged in filaments (see Figure 8.3). These filaments comprise most of the gas in molecular clouds. They have a typical width of one-tenth of a parsec and may extend for several parsecs. They precede star formation, with star formation occurring in the densest parts of the filaments.

Denser filaments are observed to be the precise locations of prestellar cores in molecular clouds; these cores are the early stage of development from which new stars may form. Sufficiently dense filaments show a chain of dense cores.

8.2.2 STELLAR JETS

Observational studies reveal the surprising result that new stars are usually associated with high velocity, highly collimated jets that may extend for long distances in the interstellar medium. A typical stellar jet originates very close to the newly forming star, within some tens of stellar radii of the star. These jets are very narrow, only a few stellar radii in diameter. Jet velocities are very high, ranging from about a hundred to about a thousand km s^{-1}. The mass loss rate associated with jets is usually in the range 10^{-6}–10^{-4} $M_{\odot}$ y^{-1}. Jets have number densities of hydrogen atoms and molecules that are usually in the range 10^{8}–10^{11} m^{-3}, and the gas in the jet is partially ionized. The

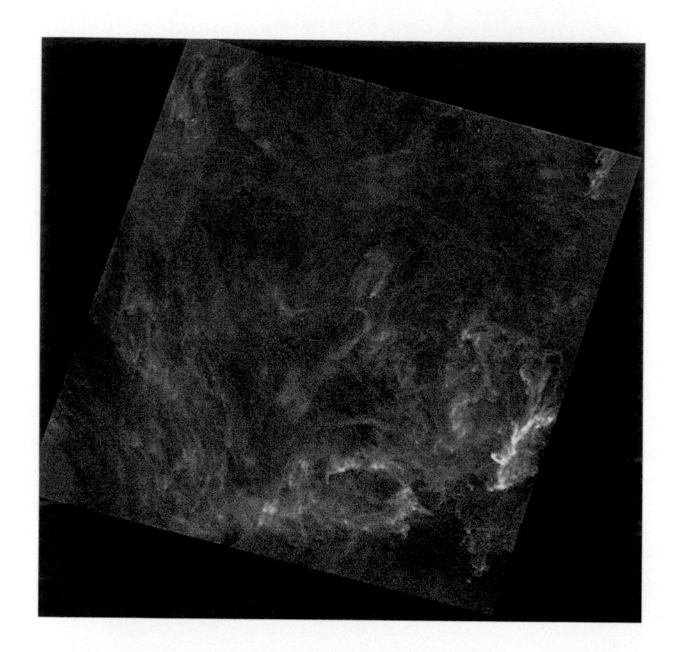

FIGURE 8.3 Image of filaments in molecular clouds in the constellation of Polaris about 160 pc distant from Earth. The image is a composite of infrared observations taken by ESA's Herschel Space Observatory at wavelengths of 250, 350, and 500 μm. (Copyright: EAS and the SPIRE and PACS consortia, Ph. André (CEA Saclay) for the Gould's Belt Survey Key Programme Consortium, and A. Abergel (IAS Orsay) for the Evolution of Interstellar Dust Key Programme Consortium.)

temperature of the jet gas is not uniform and is in the range of 10^2–10^4 K. The jet gas is also partially molecular, and contains interstellar dust. There is evidence that some of the dust is destroyed in shocks within the jet: near to the jet launch region, the fractional abundance of gaseous Fe atoms is much less than solar, while downstream their fractional abundance is near solar. This suggests that Fe-bearing dust grains are supplied at the jet's origin and then gradually eroded during the jet's evolution. An image of a jet and its counter-jet is shown in Figure 8.4.

The jets often show a 'knotty' structure in vibrational emission from molecular hydrogen and from atomic hydrogen H alpha emission. The same structure also may appear in forbidden lines from atomic ions such as sulfur, or in some molecular lines other than those from H_2. Ultimately, jets are terminated when they impact on gas either associated with the star-forming region or – if the jet escapes from that region – on interstellar gas. In the latter case, jets may be several parsecs long. The impact of the fast jet on gas results in a shock. The shock-heated gas is visible in projection against dark interstellar material as a localized region of bright emission. These localized emission regions are known as Herbig–Haro objects (or HH objects).

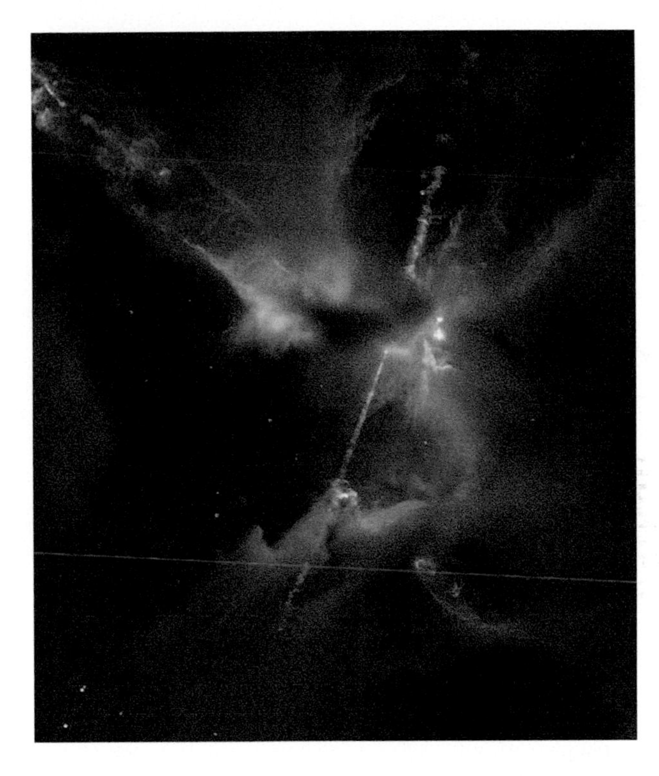

FIGURE 8.4 The central star that generates these very striking jets is hidden within a dense dusty gas core. (Credit: NASA, ESA, Hubble Heritage (STScI/AURA)/Hubble-Europe Collaboration; D Padgett (GSFC), T Megeath (University of Toledo), B Reipurth (University of Hawaii).)

8.2.3 Observations of HH Objects

As stated earlier, HH objects are formed when the passage of a stellar jet is terminated by its impact on a molecular cloud. An image of HH46/47 is shown in Figure 8.5. In this example, the newly formed star at the centre of the image generates a jet and counter-jet while variations within the velocity of the jets create shocks within the jets, making them visible. The collision of these jets with interstellar material (which is unseen in this image) creates two large bow shocks about 3 pc apart. The jets in HH46/47 travel at about 150 km s^{-1}, so the impact on the interstellar material is highly supersonic and HH objects can be important sources of visible and UV radiation within dark molecular clouds.

In fact, HH objects are sufficiently powerful radiation sources that they can drive a photochemistry in regions of molecular clouds from which radiation would otherwise be excluded. Observations show that this photochemistry produces an extensive range of specific molecular species, distinct from normal

FIGURE 8.5 Hubble image of HH46/47. (Credit: NASA, ESA, and the Hubble Heritage Team (STScI/AURA).)

cold cloud chemistry. The detection of these unexpected molecular species, generated by radiation from HH objects, allows HH objects to be used as probes of the interior structure of dark clouds.

8.2.4 Observations of Bipolar Outflows

Young stars often show bipolar outflows aligned with the stellar jets. These outflows are stellar winds that erode and entrain the ambient interstellar gas. The wide-angle bipolar outflows have velocities typically of about 30 km s^{-1}, very much slower than the jet velocities. These diverging flows inject significant amounts of mass, energy, and momentum into the interstellar medium, and erode large-scale cavities, causing mixing of wind gas with gas from the surrounding molecular cloud in the boundaries of the cavities. The consequent heating in the mixing zones of the cavities ensures that strong emission from tracer molecules such as carbon monoxide occurs at the boundaries. Injection of ionization from the wind gas into the dense, mainly neutral boundary gas, enables a rich chemistry to occur, producing tracers such as HCO^+ at higher than normal abundances.

The erosion of the ambient gas by the outflow causes the opening angle of the bipolar flows to increase with time. Initially, bipolar flows are narrow but

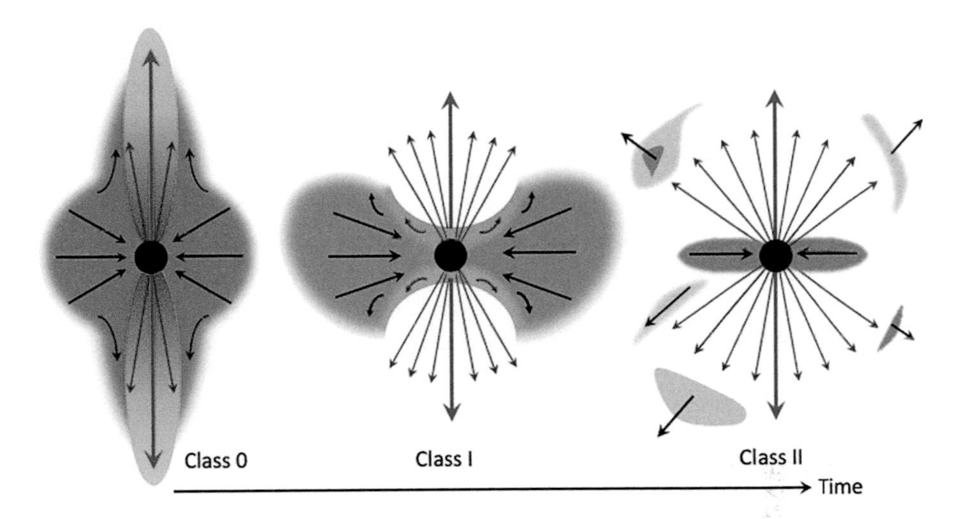

FIGURE 8.6 Evolution of bipolar flows. Class 0 refers to objects at a very early stage of evolution. Here the outflow (yellow with brown arrows) is very narrow, while infall on to the protostar (grey with black arrows) is largely unrestricted. Class I shows objects at a later stage of evolution. Here the outflow is much wider and the infall is restricted. There is an interface zone (shown in green) which has a characteristic chemistry. In Class II objects, the outflow has restricted the infall still further, the extent of the circumstellar material is very much reduced, and most of the material in the star's vicinity has been swept away by the outflow. In all classes, the bold blue arrow indicates the stellar jet and counter-jet. The image is a revised version of a figure by H G Arce and A Sargent, Astrophysical Journal (2006), 646, 1070. (Credit for the revised figure: C. Cecchi-Pestellini.)

they may eventually open wide enough to cut off any gas infalling onto the star from the surrounding cloud. An indication of this process is shown schematically in Figure 8.6.

8.3 A MODEL OF LOW-MASS STAR-FORMING REGIONS

Given the wealth of observational information that exists about low-mass star formation, what is the picture of the process that can account for these data?

The story begins with the formation of filamentary structure in molecular clouds. These filaments may be formed at the intersection of large-scale supersonic flows impacting on the molecular cloud. Each large-scale flow generates a near-planar shock, and two such flows generate a filament. The filaments evolve by accreting mass from the parent cloud so that they grow in density. Filaments with mass per unit length above a critical value become gravitationally unstable, and the local Jeans mass for the observed filaments is close to the observed peak in the prestellar core mass distribution.

There is some evidence that a switch occurs for magnetic fields originally along the filamentary structure to being perpendicular to it at a critical density of the filament; this critical density is also close to that at which gravitational instability occurs.

The next stage is the evolution of a protostar within a prestellar core: a protostar is an object that is accreting mass but in which nucleosynthesis has yet to begin. The gravitational infall of matter establishes the protostar at the centre of the core. The infall is influenced by angular momentum of the core gas, so that a circumstellar disc forms around the protostar. The circumstellar disc will eventually evolve into a protoplanetary disc from which new planets will eventually emerge. However, at this earlier star-forming stage, the infalling matter joins the circumstellar disc from which matter may be fed directly to the protostar. The magnetic field that was embedded in the filament remains in the disc, although the orientation of the field lines may be changed to be perpendicular to the disc. As the gravitational collapse proceeds, the interior of the protostar heats up enough for nucleosynthesis to begin at the centre of the object.

The material in stellar jets comes from matter infalling towards the protostar. This material is accelerated by the magnetic field fixed in the rotating disc, constrained by this rotating field, and collimated into a narrow jet. A schematic diagram of how this acceleration and collimation may occur is shown in Figure 8.7. The launch region may simply originate in the stellar wind (a few stellar radii) or it could be formed in the larger region of the disc wind (up to 30 astronomical units).

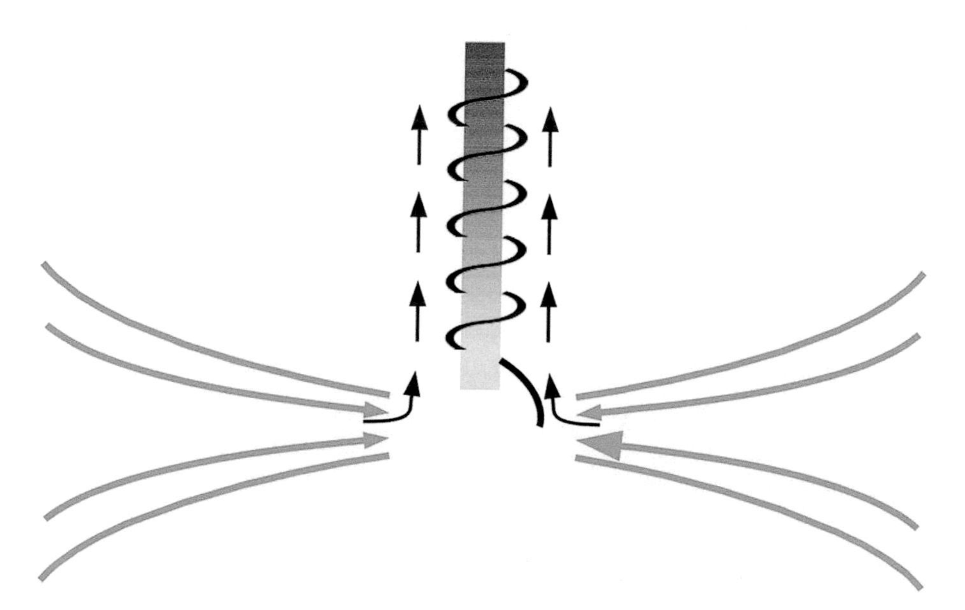

FIGURE 8.7 A schematic diagram showing how jet formation and launch may occur. Infalling matter in the rotating disc (shown in blue) is constrained by the magnetic field (in black) and confined into a narrow jet (in yellow/brown) and accelerated away from the disc. (Credit for diagram: C Cecchi-Pestellini.)

The origin of the 'knotty' structure in jets (see Section 8.2.2) is unclear and several mechanisms have been suggested. For example, periodical variations in the high jet velocity can produce an apparently 'knotty' emission in the jet, as slower velocity jet material is overtaken (and shocked) by faster material. Emission from these internal shocks may be seen as a localized clump of emission within the jet. Alternatively, it may be that the jets are not smooth, but are themselves truly clumpy in density and that these clumps have a distribution of velocities. Then slower clumps are overtaken by faster clumps, and an internal shock will occur that appears as a localized 'knot' of emission. While it is not clear which of these models is correct (or whether both are in play), it is clear that – internally – jets are not uniform and contain structure maintained by internal shocks.

Bipolar outflows accompany fast stellar jets and are readily observed because of the large cavities that they create. The outflow wind interacts powerfully with cold molecular gas at the cavity boundaries and erodes the cavity walls. In timescales on the order of 10,000 years cavities open completely, and inflow via the circumstellar disc is suppressed. Finally, the young star emerges from its birth cloud.

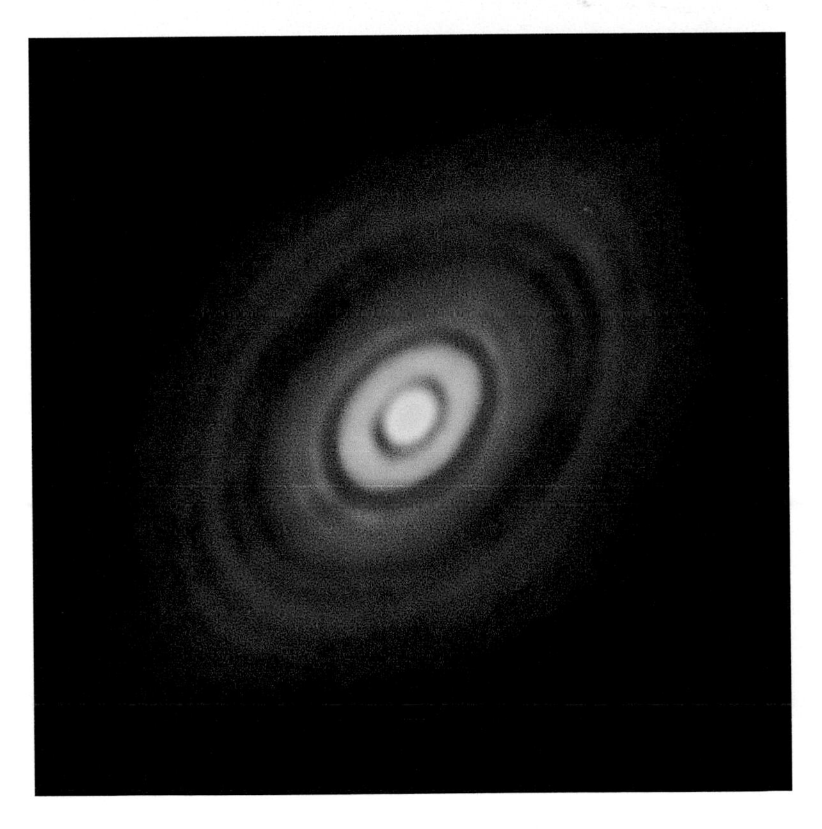

FIGURE 8.8 Image of the protoplanetary disc around the young star HL Tau. (Credit: ALMA (ESO/NAOJ/NRAO))

The circumstellar disc evolves into a protoplanetary disc of gas and dust around the newly formed star. These are low-mass objects with thickness very much less than the disc diameter. Interactions between dust grains cause the collisional growth of grains, ultimately into planetesimals from which planets may form. A false colour image of a protoplanetary disc taken by the Atacama Large Millimetre Array at submillimetre wavelengths is shown in Figure 8.8. The material in the disc appears to show gaps, indicating the formation of several planets. Figure 8.9 shows that many nearby protoplanetary discs have similar structures to that of HL Tau.

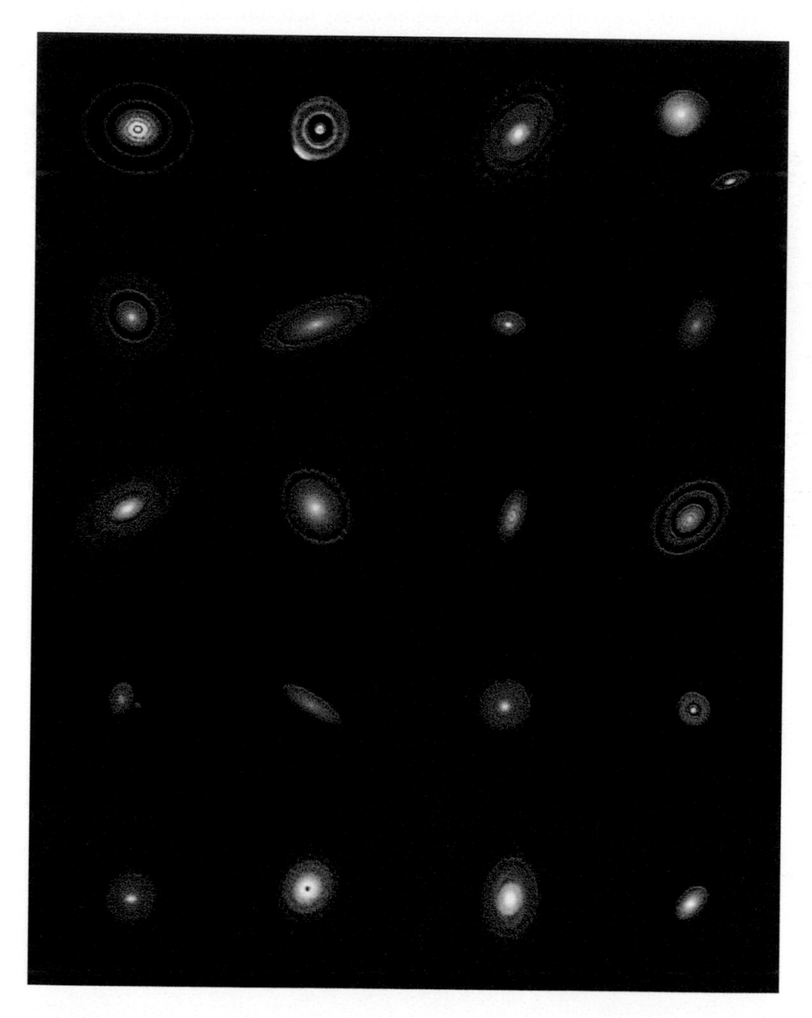

FIGURE 8.9 ALMA false-colour images of 20 nearby protoplanetary discs, showing a variety of structures. All the discs show evidence of planet formation. The implied rate of planet formation in the Milky Way galaxy appears to be high. (Credit: ALMA(ESO/NAOJ/NRAO), S.Andrews et al.; NRAO/AUI/NSF, S.Dagnello)

8.4 THE FORMATION OF HIGH-MASS STARS

From a theoretical point of view, the formation of massive stars, say of more than ten solar masses, is more complex than the formation of low-mass stars that was discussed in the previous section. For stars above a critical mass, radiation pressure from the new star on infalling dust grains is apparently powerful enough to limit the growth of the stellar mass by accretion from the surrounding gas. Thus, stars with very large masses should not exist. However, it is clear that very massive stars (even as great as 100 solar masses) are detected. Different theories to resolve this problem are currently debated, but are beyond the scope of this book.

However, observational studies of the formation of massive stars have made great progress, and a sequence of stages of massive star formation can be – at least tentatively – identified.

1. There is clear evidence that massive star formation occurs in filaments within dark molecular clouds. These structures are revealed by dust continuum emission and by line emissions from molecules such as CO and HCO^+. The filaments show infall signatures.
2. The second stage corresponds to denser cold gas that is chemically more complex. The gas is seen as dark material against the bright mid-infrared emission of the galactic background, and is therefore described as infrared dark cloud gas. However, even in this early stage, there are signatures of star-forming activity, such as emissions from shocks and masers.
3. In the third stage, a high-mass protostellar object is formed. It is still very deeply embedded in dense gas. The temperature of the gas is now increasing. The gas may be probed by emissions from a wide variety of simple molecules, formed in an active chemistry.
4. The next stage shows molecular line emissions from small (less than one-tenth of a parsec), warm (about 200 K), and dense infalling gas, close to and heated by the protostar (but distinct from it). There is an active chemistry in these small regions, forming complex organic molecules such as glycolaldehyde (CH_2OHCHO), methyl formate (CH_3OCHO), dimethyl ether (CH_3OCH_3), and methyl alcohol (C_2H_5OH). These regions are called hot cores.
5. Finally, ionized regions appear. Compared to the large HII regions discussed in Chapter 5, these are tiny regions a few percent of a parsec in diameter, and are called ultracompact ionized (UCHII) regions. They are embedded in high-density dusty gas. Outflows are also seen at this evolutionary stage. The massive star then erodes or removes all the debris from the star-forming process.

PROBLEMS

1. Calculate the free-fall collapse time of a gas cloud of radius 30 pc and mass $500 M_{\odot}$.

2. Derive an expression for the collapse time of a uniform cloud using order-of-magnitude arguments and compare it with equation (8.25).
3. At what temperature must a gas cloud of radius 0.5 pc and density 10^{10} m^{-3} be in order to avoid collapse?
4. A bubble driven by wind momentum is observed to expand at a constant velocity. Using arguments analogous to those in problem 6 of Chapter 7, find the variation of the interstellar gas density with distance from the star.

Answers to Problems

CHAPTER 2

(2) 1/2.
(4) 1.61×10^{10} Hz.
(6) 6.09 mm.

CHAPTER 3

(1) 10^4 yr.
(2) 10^7 m^{-3}.
(3) 1.3×10^{-26} J m^{-3} s^{-1}.
(4) 0.067.
(5) $n\,(AB) = n(A)[k_1 n(B) + k_2 n(BC)]/[k_3 n(D) + \beta]$.
(6) $n(H_2^+) = 10^{-2}$ m^{-3}; $n(H_3^+) = 10^2$ m^{-3}.
(7) The timescale to approach steady state is 10^5 s for H_2^+. Since 10^5 s $\ll 10^6$ years, steady state is a good approximation for the H_2^+ abundance.

CHAPTER 4

(1) 0.51 magnitudes.
(2) 4.4×10^6 yr.
(3) 2×10^9 yr.
(4) 12.6 K.
(5) 8.8×10^{24} m^{-2}.

CHAPTER 5

(1) 50.
(2) The ratio of the mass in the first cloud to that in the second is n_2/n_1. So the answer is 'no'.
(3) The average kinetic energy of a photoejected electron is

$$\langle Q \rangle = \int_{\nu_L}^{\infty} (h\nu - I_H) S_{*\nu} \mathrm{d}\nu \Big/ \int_{\nu_L}^{\infty} S_{*\nu} \mathrm{d}\nu$$

where $S_{*\nu}$ is the stellar photon production rate per unit frequency interval at frequency ν, and $h\nu_L \equiv I_H$. For a black-body at temperature T_*,

$$h\nu S_{*_\nu} \propto B_\nu(T_*) \equiv \frac{2h\nu^3}{c^2} \frac{1}{\exp(h\nu/kT_*) - 1}.$$

The integrals can be evaluated by noting that $h\nu/kT_* \gg 1$ for all frequencies in the range of integration.

(4) 2.5×10^{-3}.

(5) The wavelength corresponding to frequency ν_0 is $\lambda_0 = 1.53$ m. Hence, $\tau_{\lambda_0} \approx 10^{-5}\varepsilon$. Since $\tau_{\lambda_0} = 1$, $\varepsilon \approx 10^5$ cm^{-6} pc. If L is the path length through the nebula, $\varepsilon \approx n_e^2 L$, where n_e is the electron density. Thus, $L \approx 3 \times 10^{17}$ m. The total recombination rate within the nebula is approximately $L^3 n_e^2 \beta_2(T_e)$. Taking $T_e = 10^4$ K, $\beta_2 = 2 \times 10^{-19}$ m^3 s^{-1}. The number of UV photons generated per second by the stars (N_*) must equal the total recombination rate. Hence, $N_* \approx (3.0 \times 10^{17})^3 (10^8)^2 (2 \times 10^{-19}) \approx 6 \times 10^{49}$s^{-1}. Thus, six stars are needed.

(6) 4100 K; 11 km s^{-1}.

(7) The ionization of the envelope is governed by the Strömgren relationship, $S_* \propto \int_{r_*}^{r_u} n^2 r^2 \mathrm{d}r$, where r_* and r_u are respectively the stellar radius and the outer radius of the ionized region. Since $n \propto r^{-3/2}$, $S_* \propto \ln(r_u/r_*)$, that is, $r_u \propto e^{AS_*} r_*$ where A is a constant. If $AS_* \ll 1$, $r_u \approx r_*$; if $\gg 1, r_u \gg r_*$.

CHAPTER 6

(2) 2×10^5 yr.

(4) The Rankine–Hugoniot conditions for a strong shock propagating in a gas of $\gamma = 7/5$ are

$$P_2 + \rho_2 u_2^2 = P_1 + \rho_1 u_1^2 \approx \rho_1 u_1^2$$

$$\rho_2 u_2 = \rho_1 u_1$$

$$\frac{1}{2}u_2^2 + \frac{7}{2}\frac{P_2}{\rho_2} = \frac{1}{2}u_1^2 + \frac{7}{2}\frac{P_1}{\rho_1} \approx \frac{1}{2}u_1^2.$$

Hence,

$$\left(\frac{\rho_2}{\rho_1}\right)^2 - 7\left(\frac{\rho_2}{\rho_1}\right) + 6 = 0$$

that is,

$$\frac{\rho_2}{\rho_1} = 6.$$

(5) From equation (5.34) with typical values $n_e = 10^8 \text{m}^{-3}$, $T_e = 10^4\text{K}$, $\mathcal{L}_{O^+} \approx 2.2 \times 10^{-21}\text{J m}^{-3}\text{ s}^{-1}$. The thermal energy content per unit volume is $3n_e kT_e \approx 4.1 \times 10^{-11}\text{J m}^{-3}$. The cooling time $t_c \approx 3n_e kT_e/\mathcal{L}_{O^+} \approx 1.8 \times 10^{10}$ s. If the shocked gas has velocity $\approx$ 10 km s^{-1}, the cooling length $L_c \approx 1.8 \times 10^{14}$m. This is very much less than the radius of the HII region.

(6) $n_s = 4 \times 10^7\text{m}^{-3}$; $T_s = 2.7 \times 10^6 K$. The velocities in the shock and fixed fames are respectively 125 km s^{-1} and 375 km s^{-1}, respectively. The cooled gas has velocity 500 km s^{-1} in the fixed frame and is at rest relative to the shock.

(7) $t_0 = 2.6 \times 10^{12}$s; $l_c = 6.5 \times 10^{17}$m.

CHAPTER 7

(1) 1:10.

(3) From equations (7.41) and (7.42), the outer shock radius and velocity after 30 000 yr are respectively $R_s \approx 6.6 \times 10^{16} m$ and $V_s \approx 43\ kms^{-1}$. The Mach number of the shock is therefore approximately 3.3. From equation (6.80), the compression ratio across the shock is about $(3.3)^2 \approx 11$. Thus, the shocked gas has a density 1.1×10^8 m^{-3}. From mass conservation, therefore, the thickness of the shocked region is $\Delta R_S \approx (1/3)(10^7/1.1 \times 10^8)R_s \approx 2.2 \times 10^{15}$m.

(4) The average pressure in the bubble, P say, is proportional to the thermal energy density. Hence, $P \propto R^{-3}$ where R is the radius of the bubble (i.e. the shock radius). If the upstream density is $\rho(R)$, then $P \propto \rho(R)\,\dot{R}^2$. Since $\rho(R) \propto R^{-2}$, $P \propto \dot{R}^2 R^{-2}$. Hence, $\dot{R}^2 \propto R^{-1}$ and $R \propto t^{2/3}$, $\dot{R} \propto t^{-1/3}$.

(5) At time t, the bubble has radius $R(t)$ and boundary velocity $V(t)$ given by equations (7.66) and (7.67), respectively. The gas temperature is $T(t) = 3m_H V(t)^2/32k$ and the density is $n(t) = 4n_0$, where n_0 is the interstellar gas density. At this time, the bubble loses energy by free–free radiation at a rate

$$\mathcal{L}(t) \approx \frac{4}{3}\pi R(t)^3 \mathcal{L}_{\text{ff}}(t)$$

where

$$\mathcal{L}_{\text{ff}}(t) \propto n(t)^2 T(t)^{1/2} \propto n_0^2 V(t).$$

The total energy radiated away up to time t is given by

$$\mathcal{L}_{\text{Tot}}(t) = \int_0^t \mathcal{L}(t')\text{d}t'$$

$$\propto \int_0^t n_0^2 V(t') R(t')^3 \mathrm{d}t'.$$

Evaluation of this integral at $t = 10^4$ yr for $E_* = 10^{44}$ J and $n_0 = 10^6$ m^{-3} gives $\mathcal{L}_{Tot} \approx 2.8 \times 10^{41}$J. Hence, about 0.3% of the explosion energy has been radiated away.

(6) Let the density around the star be $\rho = \rho_* \, r^{-\beta}$. The appropriate density parameter is ρ_*, which has dimensions $[M]^1[L]^{\beta-3}$. Assume $R \propto \dot{E}_*^{\alpha} \rho_*^{\gamma} t^{\lambda}$ and obtain $\alpha = \frac{1}{3}, \gamma = -\frac{1}{3}, \lambda = 1$ so $\beta = 2$.

(7) The supernova affects its surroundings up to a time when its interior pressure is about equal to the pressure of its surroundings. This happens at an expansion velocity $V_s \approx 4$ km s^{-1}. This is in the snowplough phase. Using the snowplough equations with $V_0 = 250$ km s^{-1} and $R_0 = 7 \times 10^{17}$ m gives a time $t \simeq 1.7 \times 10^{14}$ s, radius $R \approx 2.8 \times 10^{18}$ m and a mass affected $M \approx 10^5 M_\odot$.

CHAPTER 8

(1) 1.2×10^8 yr.

(2) Consider a mass on the surface of the cloud. The initial cloud radius is R and initial density ρ_0. The initial acceleration of the mass is therefore given by

$$\ddot{R} = \frac{4}{3}\pi G R \rho_0.$$

Replacing $\ddot{R}$ by R/t^2, where t is the collapse time, we find

$$t \approx \sqrt{3/4\pi G \rho_0}.$$

(3) 27 K.

(4) $\rho \propto r^{-2}$.

Index

A

B

C

D

E

F

G

H

I

J

K

L

M

T

U

V

W

X

Z